MOUNTAIN RESEARCH IN EUROPE

An Overview of MAB Research from the Pyrenees to Siberia

MAN AND THE BIOSPHERE SERIES

MAN AND THE BIOSPHERE SERIES

Series Editor J.N.R. Jeffers

VOLUME 14

MOUNTAIN RESEARCH IN EUROPE

An Overview of MAB Research from the Pyrenees to Siberia

M. F. Price

Environmental Change Unit
University of Oxford, Oxford, UK

PUBLISHED BY

UNESCO

PARIS

AND

The Parthenon Publishing Group

International Publishers in Science, Technology & Education

Published in 1995 by the United Nations Educational, Scientific and Cultural Organization,
7 Place de Fontenoy, 75700 Paris, France — UNESCO ISBN 92-3-103050-7

and

The Parthenon Publishing Group Limited
Casterton Hall, Carnforth
Lancs LA6 2LA, UK—ISBN 1-85070-570-4

and

The Parthenon Publishing Group Inc.
One Blue Hill Plaza
PO Box 1564, Pearl River,
New York 10965, USA—ISBN 1-85070-570-4

Typeset by Lasertext Ltd., Stretford, Manchester
Printed and bound in Great Britain by
Butler and Tanner Ltd., Frome and London

British Library Cataloguing in Publication Data

Price, Martin F.
Mountain Research in Europe: Overview of MAB Research from the Pyrenees to Siberia — (Man & the Biosphere Series; Vol. 14)
I. Title II. Series
574.5264

ISBN 1-85070-570-4

Library of Congress Cataloguing-in-Publication Data

Price, Martin F.
Mountain research in Europe: an overview of MAB research from the Pyrenees to Siberia/M.F. Price.
p. cm. — (Man and the biosphere series; v. 14)
Includes bibliographical references and index.
ISBN 1-85070-570-4 (Parthenon Pub. Group)
1. Man—Influence on nature—Europe. 2. Mountain ecology—Europe. 3. Unesco. Programme on Man and the Biosphere I. Title II. Series
GF540.P75 1995
304.2′0914′3—dc20 94-27665
CIP

PREFACE

UNESCO'S Man and the Biosphere Programme

Improved scientific understanding of natural and social processes associated with the environment is the focus of UNESCO's Man and the Biosphere Programme. It works towards this aim through the provision of information useful to resource managers and administrators and by promoting the conservation of genetic diversity. Much work is devoted to problem-solving ventures that bring together scientists, policy-makers and local people. In parallel with this, the programme seeks to develop scientific networks and strengthen regional co-operation.

Launched in the early 1970s, the programme is nationally based but internationally co-ordinated. MAB emphasizes problem solving research, training and demonstration. Often interdisciplinary teams are assembled to analyse interactions between ecological and social systems. A systems approach is used to understand relationships between development and the environment.

MAB is a decentralized programme with field projects and training activities in all regions of the world. Universities, academies of sciences, national research laboratories and other research and development institutions are all involved and brought together under the auspices of more than 100 MAB National Committees. Co-operation with a range of international government and non-governmental organizations is characteristic of the MAB Programme.

Man and the Biosphere Book Series

The Man and the Biosphere Book Series was launched to help communicate results from the MAB Programme. Primarily aimed at upper level university students, scientists and resource managers, the books are not normally suitable as undergraduate textbooks but rather seek to provide additional resource material. Some books are based on primary case studies that involve original data collection. Others provide a synthesis of global and regional research across several sites or countries. In all cases, a state of the art assessment of knowledge and methodological approaches is sought. In some cases, books are commissioned. The series editor is John Jeffers, formerly Director of the Institute of Terrestrial Ecology in the United Kingdom, who has been associated with MAB since its inception.

Mountain Research in Europe

This volume presents an overview of scientific activities on mountain systems undertaken within the framework of MAB in the broad European region, from the Pyrenees in the west to Siberia in the east. MAB work on temperate mountains has been largely planned and carried out within the framework of one of the fourteen international project areas set up within the programme – MAB Project 6 (MAB-6) on 'Impact of human activities on mountain and tundra ecosystems'.

The main outlines of MAB Project 6 were sketched out in 1973 at international meetings held at Salzburg (Austria) and Lillehammer (Norway), with different research themes being proposed for particular regions (i.e. tropical high-mountain areas, high-latitude mountain and tundra ecosystems, and temperate high mountains), as well as topics which could be addressed in worldwide comparative research (e.g. delineation of regions subject to natural hazard, identification of environmental controls and human impact on world timberlines).

In the mid-latitude temperate zones, the primary focus was on tourism, technology and land-use alternatives in mountain systems with two distinct seasons, with an emphasis on the impact of summer and winter tourist and recreational activities and on the massive modification of areas left abandoned through cessation of traditional land-use practices. An initial study of the small Alpine community of Obergurgl in Austria acted as a forerunner of a network of projects in the European Alps, with studies in other mountain regions of the broader European region being shaped during the late 1970s and 1980s by MAB National Committees and collaborating research institutions.

Providing an overview of this work within MAB on mountain research in Europe is the aim of the present volume, written and compiled by Martin Price. The overview is based on the analysis of published and unpublished documents, and interviews and field visits in 1992 and 1993 in the countries covered by the overview: Austria, France, Germany, Spain and Switzerland in western Europe; Bulgaria, Czech Republic and Slovakia, Poland and the former Soviet Union in central and eastern Europe. For each country, the analysis follows a similar structure, addressing the mountain regions of the country, the organization of MAB-6 activities, MAB-6 research activities and conclusions. A general introduction to the volume sets the background to MAB work in mountain areas in Europe, with a closing chapter providing an overall evaluation of that work in terms of the use of models, organizational and institutional issues, lessons for interdisciplinary research, and legacy for the future.

The volume thus covers organizational and institutional as well as substantive aspects of mountain research in Europe within the framework of MAB. As such, it is hoped that the book will be of interest to those concerned with the planning of collaborative research initiatives at national, regional and international levels, as well as more substantively with those who study mountain environments and communities and who shape, move and implement policies which affect mountain land use and resource management.

In this vein, it is hoped that some of the approaches and results described in this book will be considered by readers as being in tune with the spirit of a Russian proverb, cited a quarter-of-a-century ago in a book to mark the twentieth anniversary of the Universal Declaration of Human Rights (*Birthright of Man*, UNESCO 1969): 'To be keen-eyed is to see, not the mountain, but what lies behind it'.

UNESCO expresses its thanks to Professor Jack Ives for contributing the foreword to this volume and to Dr Martin Price for having carried out the overview exercise and for seeing this volume into print. Thanks are also extended to those many institutions and individuals, acknowledged in the Author's Preface, who provided hospitality, information and insights during the author's visits to their countries, and who reviewed draft chapters and contributed in other ways to this volume.

MAN AND THE BIOSPHERE SERIES

---

1. The Control of Eutrophication of Lakes and Reservoirs. S.-O. Ryding; W. Rast (eds.), 1989.
2. An Amazonian Rain Forest. The Structure and Function of a Nutrient Stressed Ecosystem and the Impact of Slash-and-Burn Agriculture. C.F. Jordan, 1989.
3. Exploiting the Tropical Rain Forest: An Account of Pulpwood Logging in Papua New Guinea. D. Lamb, 1990.
4. The Ecology and Management of Aquatic-Terrestrial Ecotones. R.J. Naiman; H. Décamps (eds.), 1990.
5. Sustainable Development and Environmental Management of Small Islands. W. Beller; P. d'Ayala; P. Hein (eds.), 1990.
6. Rain Forest Regeneration and Management. A. Gómez-Pompa; T.C. Whitmore; M. Hadley (eds.), 1991.
7. Reproductive Ecology of Tropical Forest Plants. K. Bawa; M. Hadley (eds.), 1990.
8. Biohistory: The Interplay between Human Society and the Biosphere – Past and Present. S. Boyden, 1992.
9. Sustainable Investment and Resource Use: Equity, Environmental Integrity and Economic Efficiency. M.D. Young, 1992.
10. Shifting Agriculture and Sustainable Development: An Interdisciplinary Study from North-Eastern India. P.S. Ramakrishnan, 1992.
11. Decision Support Systems for the Management of Grazing Lands: Emerging Issues. J.W. Stuth; B.G. Lyons (eds.), 1993.
12. The World's Savannas: Economic Driving Forces, Ecological Constraints and Policy Options for Sustainable Land Use. M.D. Young; O.T. Solbrig (eds.), 1993.
13. Tropical Forests, People and Food: Biocultural Interactions and Applications to Development. C.M. Hladik; A. Hladik; O.F. Linares; H. Pagezy; A. Semple; M. Hadley (eds.), 1993.

CONTENTS

AUTHOR'S PREFACE AND ACKNOWLEDGEMENTS

This book is based mainly on visits to the nine (now ten) countries under review between July 1992 and April 1993. However, it has had a long evolution, as it brings together two primary emphases of my work since the late 1970s: first, resource management in mountain regions and, second, the organization and conduct of effective applied interdisciplinary research. My first experience in both of these areas was during my M.Sc. at the University of Calgary, supervised by Allan Legge, to whom I owe a great deal for his support and advice over the years. At my thesis defence in 1982, Walter Moser, the former co-ordinator of the MAB-6 project in Obergurgl, Austria, was my external examiner. When we next met 2 years later, he strongly supported my idea of a comparative study of resource management in the Alps and the Rockies, and provided introductions to Bruno Messerli, who gave me a friendly and positive reception when I visited him in Berne a few months later. Shortly afterwards, I met Jack Ives at the University of Colorado at Boulder, and we agreed that I should study for a Ph.D. with him.

My initial Ph.D. concept was to undertake a comparative study of MAB-6 projects in western Europe and North America. This was, however, rather over-ambitious and premature, as many of the projects were still underway and the linguistic difficulties would have been substantial. Consequently, my Ph.D. focused only on forests in Switzerland and Colorado, including 8 months in 1985–1986 working within the Swiss MAB programme in the Geographical Institute of the University of Berne. I would like to express many thanks to Bruno Messerli and Paul Messerli, both former co-ordinators of the Swiss MAB programme, for their heartfelt welcome, patience, and support in 1985–1986 and on many subsequent visits. I would also like to thank Jack Ives for his thoughtful support during and since my Ph.D., and for his foreword to this book.

Throughout the late 1980s, while working primarily on the human aspects of climate change, based in Colorado, I maintained contact with Thomas Schaaf, of the MAB Secretariat in Paris, and we attempted to re-invigorate MAB mountain activities with an ambitious initiative on 'The sustainable future of mountain communities: resources and tourism in the context of climate variability and change'. Unfortunately, although MAB's International Co-ordinating Council approved the initiative in 1990, no financial support materialized. Nevertheless, I hope that some of the ideas may yet be acted on, particularly through the implementation of Chapter 13 on 'Managing fragile ecosystems: sustainable mountain development' of *Agenda 21*, the plan for action approved at the Earth Summit in Rio de Janeiro in 1992.

Soon after I returned to Europe in 1991, Thomas Schaaf and I revived the concept of a synthesis of MAB-6 projects. This time, the idea was more timely, as most projects were complete, and there were other calls for evaluation. As a result, UNESCO commissioned me to prepare an overview of the projects in central and eastern Europe and the former USSR in 1992, and those in western Europe in 1993. I would like to acknowledge the considerable help I have received from the MAB Secretariat then and since: not only from Thomas Schaaf, but also from Malcolm Hadley, who provided much useful advice as I turned my initial reports into this book.

My first experience of the mountains of central Europe had been in 1978, when I visited the Slovak Tatras. However, although I had seen evidence of the significant scientific activities underway there, I had never been able to obtain much information about this research. In spring 1991, I revisited the Tatras, and was warmly welcomed by Peter Fleischer and his colleagues on the scientific staff of the Tatra National Park. I continued west to the Giant Mountains, where I participated in a meeting of the Czechoslovak Biosphere Reserve Committee at the invitation of Jan Jeník of the Department of Botany at Charles University, Prague. I owe him, and Květa, much for their hospitality and willingness to explain the complexities of science under communist regimes. Czechoslovakia was also the first stop on the journeys that resulted in this book, and I would also like to thank Vera Hruškova (Czech National MAB Secretariat, Prague) and Jan Stursa (Krkonose National Park Administration, Vrchlabi) for their help. From Czechoslovakia, I continued to Poland, Russia, and Bulgaria during summer and early autumn 1992.

In Poland, I would like to thank especially Alicja Breymeyer, Chair of the Polish MAB Committee, for her support and for organizing my itinerary. Others who helped me with information and in other ways include Maria Goczot-Gontarek, Maciej Gontarek, Andrzej Raj, Ryszard Mochola, and Jozef Szafran (Karkonosze National Park); Zbigniew Krzan, Marek Kot and Pawel Skawinski (Tatra National Park); Halina Piekos-Mirkowa (Polish Academy of Sciences' [PAS] Tatra Scientific Station, Zakopane);

and, in Krakow, Adam Kotarba (PAS Institute of Geography and Spatial Organization), Zbigniew Mirek (W. Szafer Institute of Botany), Zygmunt Denisiuk and Tadeusz Zajac (PAS Nature Conservation Research Centre).

In Russia, my colleagues at the International Mountain Research and Information Laboratory in Moscow spent a hectic week sifting through a mountain of documents with me. Yuri Badenkov, Olga Galtseva, and Olga Salnikova deserve special thanks for their patience with my minimal Russian and for memorable cultural occasions on this and previous visits to Moscow and the mountains of the former USSR. Subsequently, Valeri Neronov, Deputy Chairman of the Russian MAB Committee, provided valuable comments on the evolution of activities in his country; Viktor Hawrylenko (National MAB Committee, Kiev) and Ludmila Malysheva (Faculty of Geography, Kiev University) updated me on activities in Ukraine.

My time in Bulgaria was well organized by Boyan Manev and Rumiana Zafirova of Sunshine Tours, Sofia. They and Peter Petrov (Bulgarian Academy of Sciences' [BAS] Institute of Geography, Sofia) arranged my visits to a number of biosphere reserves, with the help of Lazar Bechev (Tsarichina) and Kocho Kochev and Dimitar Petkov (Steneto). They also facilitated my meetings with key scientists in Sofia, including Alexander Alexandrov and Roumen Milanov (BAA Institute of Forestry Research); Marin Bachvarov and Nikolina Popova (Department of Geography, Sofia University 'Kliment Ohridski'); Simeon Nedialkov (Faculty of Ecology, Varna University); Dimitar Peev and Stefan Kozuharov (BAS Institute of Botany); Ivan Raev (Committee of Forests, Council of Ministers); Zheko Spiridonov (Ministry of Environment, Sofia); and Yordan Uzunov, Margarita Poromanska, and Danka Tankova (BAS Institute of Ecology).

I spent much of winter/spring 1993 updating my earlier experience of MAB-6 in western Europe, and also learning about significant projects of which I had little previous knowledge. Catherine Cibien and Jacques Lecomte of the French MAB Secretariat in Paris provided all the necessary documents and answered many questions on the complexities of MAB in the mountains of France. In Germany, the wealth of experience accumulated during the Berchtesgaden project was shared and analysed in meetings with Jörg Schaller and Michael Ashdown (Planungsbüro Dr Jörg Schaller, Kranzberg); Lutz Spandau (Allianz Stiftung, Munich); and Hubert Zierl and Werner d'Oleire-Oltmanns (Nationalpark Berchtesgaden). Karl-Heinz Erdmann, of the MAB Secretariat in Bonn, provided further institutional details.

Meetings across Austria provided many fascinating insights into the various Austrian projects, from Hans Adam and John Haslett at the Institute of Zoology, University of Salzburg; Friedrich Ehrendorfer (Institute of Botany) and Georg Grabherr (Institute of Phytophysiology) at the University of Vienna; Helmut Beissmann (Institute of Cartography and

Gerlinde Schrammel (International Programmes) at the Austrian Academy of Sciences, Vienna; and Alexander Cernusca (Institute of Botany) and Gernot Patzelt (Institute for High Mountain Research) at the University of Innsbruck. I would also like to thank two former members of the University of Innsbruck for their assistance: Christian Körner, now at the Institute of Botany, University of Basel, Switzerland; and Walter Moser, who kindly provided some useful comments from retirement in Tofield, Alberta, Canada.

Visits to Spain, building on introductions from Cristina Herrero, of the MAB Secretariat, in Madrid, introduced me to previously unknown mountains and MAB projects. I would like to thank Enrique Balcells Rocamora, Bernardo Alvera, and Federico Filla Estaqué (Instituto Pirenaico de Ecologia, Jaca) for their information about research in Aragon; and Ramon Ganyet Solé, Xavier Campillo i Besses, Albert Villaró Boix (Ajuntament de La Seu d'Urgell), and Antoni Tulla Pujol (Department of Geography, Autonomous University of Barcelona) for introducing me to the Catalan Pyrenees.

The appropriate last stop on my travels for this book was the Institute of Geography at the University of Berne, where I renewed many old acquaintances. I would especially like to thank Urs Wiesmann, Margret Möhl, and Markus Wyss at the institute, as well as Erwin Stucki (Centre for Rural, Mountain, and Environmental Studies, Chateau d'Oex) and Otto Wildi (Federal Institute for Forestry Research, Birmensdorf), for their assistance. Finally, I am grateful to Martin Parry, former Director of the Environmental Change Unit at the University of Oxford, for providing essential institutional support since I arrived here in October 1992.

The friends and colleagues mentioned above not only provided hospitality, information, and insight during my visits, but also reviewed draft chapters and provided many of the figures which I have chosen to provide a flavour of MAB research in the European mountains. I could not have written this book without their willing assistance, and hope that they find my assessment to be fair and interesting. As the MAB programme in its present form winds down, I hope this book will reveal the great wealth of experience that it has catalysed in mountain ranges from the Pyrenees to Siberia, and may provide a useful starting point for understanding and fruitful interactions between those who live and conduct scientific research in the mountains.

Martin F. Price
Oxford, April 1994

LIST OF ABBREVIATIONS

ADPE	Association for the Development of the Pays d'Enhaut (Switzerland)
AUB	Autonomous University of Barcelona
BAS	Bulgarian Academy of Sciences
CEFE	Centre d'Ecologie Fonctionelle et Evolutive (Montpellier, France)
CEHEM	Centre for the Management and Restoration of Disturbed Ecosystems in Hercynian Mountains (Czech Republic)
CEMAGREF	Centre national du Machinisme agricole, du génie rural et des Eaux et Fôrets (France)
CEPE	Centre d'Etudes phytosociologiques et écologiques (Montpellier, France)
CERME	Centre for Rural, Mountain and Environmental Studies (Switzerland)
CIPRA	International Committee for the Protection of the Alps
CIS	Commonwealth of Independent States
CNRS	Centre National de la Recherche Scientifique (France)
CPBE	Pyrenean Centre for Experimental Biology (Spain)
CSIC	Consejo Superior de Investigaciones Cientificas (Spain)
DGRST	Délégation Générale à la Recherche Scientifique et Technique (France)
DPMPE	Direction du Projet MAB-Pays d'Enhaut (Switzerland)
EAFV	Federal Institute for Forestry Research (Switzerland)
EHESS	Schools of Further Education in Social Sciences (France)
ENSAM	National High School of Agronomy (France)
ETH	Federal Polytechnic University (Switzerland)
EuroMAB	European MAB Committees
FOEP	Federal Office for Environmental Protection (Switzerland)
GAP	Grups de l'Alt Pirineu (Spain)

GIS	Geographic information system
IBP	International Biological Programme (ICSU)
ICALPE	International Centre for Alpine Environments (Le Bourget-du-Lac, France)
ICC	International Co-ordinating Council (MAB)
ICSU	International Council of Scientific Unions
IGBP	International Geosphere–Biosphere Programme (ICSU)
IIASA	International Institute for Applied Systems Analysis (Laxenburg, Austria)
IMRIL	International Mountain Research and Information Laboratory (Russian Federation)
INA	Institut National d'Agronomie (France)
INRA	Institut National de la Recherche Agronomique (France)
IPE	Instituto Pirenaico de Ecologia (Spain)
IRE	Institute of Rural Economy (Switzerland)
IUCN	World Conservation Union
MAB	Man and the Biosphere Programme (MAB)
MOPU	Ministry of Public Works and Urbanization (Spain)
MPEA	MAB-Pays d'Enhaut Association (Switzerland)
NCCEEP	National Co-ordination Centre for Ecology and Environmental Problems (Bulgaria)
PAS	Polish Academy of Sciences
PIREN	Programme interdisciplinaire de recherche sur l'environnement (CNRS, France)
PLA	Protected Landscape Area
RESALP	Réseau international de documentation et d'information de la montagne alpine
RNED	National Network for Experimentation and Development (France)
SCOPE	Scientific Committee on Problems of the Environment (ICSU)
SNF	Swiss National Foundation
SRETIE	Service de La Recherche, des Etudes et du traitement de l'Information sur l'Environnement
TANAP	Tatra National Park (Slovakia)
UNU	United Nations University (Tokyo)

FOREWORD

Early one morning in the autumn of 1972 in my Boulder, Colorado office I received a long-distance telephone call that effectively altered my life and, as this book will imply, the lives of many others whose scholarly interests involved mountain environments. I listened attentively to the voice of Francesco di Castri from UNESCO, Paris. I learned that UNESCO had initiated a fourteen-phase programme, in a sense a follow-on to the International Biological Programme (IBP), but one involving the interrelationships between human activities and the biosphere. One of the fourteen, MAB Project 6, was entitled: *The impact of human activities on mountain ecosystems*. It was to be launched through a 'panel of experts' meeting in Salzburg, Austria in January–early February, 1973. There would be about twenty-five invited scholars from around the world, and a field excursion to Obergurgl in the Tyrol. Could I attend? How could I refuse? And I was in a position to extend my stay in the Alps to undertake a month's personal tour of research institutions and to make numerous contacts with colleagues in Austria and Switzerland.

I make no apologies for a rather personal approach to this foreword. I think that those elements of what became MAB-6 research that bore the greatest success did so because of the evolution of an expanding series of highly personal relationships, many of which have continued to this day. Much of this derived from the inspired leadership of Dr Francesco di Castri and his team in the UNESCO MAB Secretariat. This assured MAB-6 not only of a range of activities from the Pyrenees to Siberia, but throughout the mountains of the world.

One of the components of long-term research that is so often lost or neglected is the overview, evaluation and assessment. This is especially true of international and interdisciplinary programmes that, by their nature or design, have no strong centralized directing force. Martin Price and UNESCO have provided an invaluable service by ensuring that MAB-6 – or at least the European and former Soviet section of it – has been afforded this treatment. The report is comprehensive and highly competent. It demonstrates the enormous progress in interdisciplinary and multi-disciplinary mountain research over the two decades. It also demonstrates differences in approach between the Eastern and Western sectors of the mountain axis, as well as between individual country, and within-country, programmes. It provides much needed bibliographic information. It indicates evolution of research design over time.

The instances of relative success and those of lack of attainment of potential are made clear, yet it says little about the personalities involved. While I am subject to a personal bias from not having equal access to all of the major players, I will try to offset some of the objectivity of Martin Price's text in this foreword.

I accepted the invitation to write this foreword with great pleasure, an honour if you will, but also a duty. The act of writing has produced waves of nostalgia as it evokes many precious moments in the mountains with old friends, development of admiration for the mountain people who have interacted courageously with their environment for centuries, frozen noses and hot leeches, and schnapps on the midnight summit of Hohe Nebelkogl with Walter Moser, the Obergurgl Model laid out at our feet.

MAB Project 6 grew out of Salzburg and Obergurgl, winter 1973, and the follow-on International Working Group meeting held later the same year in Lillehammer, Norway. The Salzburg 'panel of experts' set the stage and introduced us to the Obergurgl Model, then in its infancy and derived from IBP and the ingenuity of Walter Moser. It also introduced us to each other, initially a small group, deliberately selected to cross the human sciences and the natural sciences. The group included Paul T. Baker, Corneille Jest, Heinz Franz, Bruno Messerli, Walter Moser, Ray Dasmann, Olof Hedberg, O.V. Makeev, Luis Briançon, M. Numata, and Paul Veyret, our Salzburg hosts, the UNESCO MAB Secretariat, especially Michel Batisse, Francesco di Castri and Gisbert Glaser, and, as keynote speaker, Konrad Lorenz. Little did we realize that the striking simplicity of Walter Moser's Obergurgl Model (Mark I) would provide a corner stone and avenue to considerable sophistication for years to come.

Lillehammer, the following November, was on the outside sombre and dark, in accordance with the latitude, and on the inside it produced a level of exchange, activity, and drafting of documents that amaze me to this day. The group was an enlargement on the Salzburg 'panel of experts' and made more vigorous attempts to integrate the human and natural sciences (Kenneth Hewitt, David C. Pitt, Eilif Dahl, Bill Heal, Frans Wielgolaski, Fred Bunnell, Olivier Dolfuss, Paul Ozenda, Heinz Ellenberg, M.S. Mani, Bihari K. Shrestha, and Olaf Rønning, amongst others, were added). In addition there was a corps of IBP Tundra Biome modellers. I had the dubious privilege of being 'elected' chairperson of the International MAB-6 Working Group and, pressed by di Castri and Malcolm Hadley, found myself writing and rewriting drafts of what became parts of UNESCO MAB Lillehammer Report, No. 14, until 2 a.m. The two documents from Salzburg (MAB Rept. No. 8) and Lillehammer became the central reference for most of the future development of MAB research, only to be rendered obsolete by the increasing sophistication of Swiss MAB-6 some years later.

There followed regional MAB meetings in Vienna, Munich, La Paz, the Briançonnais, Kathmandu, Boulder, Lima, Bogota, and elsewhere. Many of the original Salzburg group attended all or most of these meetings. The networks developed and enlarged and the MAB Secretariat members in Paris performed the superhuman task of editing and publishing not only MAB-6 reports, but reports from all fourteen MAB projects.

From my perspective, MAB-6 Salzburg–Lillehammer was a flashpoint of collective mountain scholarship. It not only set on its course a large assemblage of independent national MAB programmes, it permeated the International Geographical Union's Commission on Mountain Geoecology, it influenced the establishment of the International Centre for Integrated Mountain Development (ICIMOD) in Kathmandu, its focus was transferred into the United Nations University's mountain programme, it lapped over into the establishment of *Mountain Research and Development* and the founding of the International Mountain Society. The spurt of mountain research reached the special international Study Weeks of the Vatican, and its impetus aided the efforts of Mountain Agenda-92 to push for the inclusion of a specific chapter on Mountains (Chapter 13) in the UNCED (Earth Summit) Rio de Janeiro document, known as Agenda 21.

MAB Project 6 began with extraordinary ideals and a level of ambition that could not possibly be sustained. One could argue that the significant level of success could have been much greater had there been more funds and a more centralized control. In retrospect it is clear that this argument is unsound. The very spontaneity of many of the national MAB-6 programmes indicates that the brain-child of Michel Batisse, the entire MAB Programme, was correctly conceived. Some general lessons, however, are worthy of emphasis. Interdisciplinary research is a rarity and is extraordinarily difficult to achieve. In that MAB-6 went part of the way, a ranking of fair-to-good, with some excellent scores and some disappointments, is due. It is especially difficult for human scientists and natural scientists to work in a fully integrated fashion. It is even more difficult for such research to penetrate the decision-making process. Nevertheless, remarkable progress was made in all these areas and there is a much increased awareness of the mountain problematique than existed in 1973. Just as the eruption of ecosystem modelling with IBP, along with euphoria, disappointments with the realization that modelling is a tool not a panacea, so the much greater complexity upon introduction of the human being into the ecosystem must be appraised in perspective. And the very euphony of *Man and the Biosphere* indicates that the programme arose prior to the acknowledgement that more than half the mountain subsistence farmers of the world are female.

There is no doubt that not only are there MAB-6 loose ends and unfinished business, but that MAB-6 type research will go on for a very long time. I am sure that the efforts of Martin Price, in undertaking this assessment of 20 years of MAB-6 work, will prove an essential source of reference. It remains for UNESCO to complete the appraisal of MAB-6 research that has extended into practically all sections of the mountain world.

Jack D. Ives 15 March, 1994

Former Chair, International Working Group, UNESCO MAB-6;
President, International Mountain Society;
Co-ordinator, United Nations University Programme on Mountain Ecology and Sustainable Development;
Professor, University of California, Davis, CA, USA

CHAPTER 1

GENERAL INTRODUCTION

Throughout the 1970s and 1980s, UNESCO's Man and the Biosphere (MAB) programme was the major international interdisciplinary research programme involving natural and social scientists. In many ways, it may be regarded as a successor to the 1964–1974 International Biological Programme (IBP) of the International Council of Scientific Unions (ICSU) (Worthington, 1975), which led to major advances in biology and ecology. The initial concept of the MAB programme came from the participants in the 'Biosphere Conference', held in Paris in 1968, who recognized the need for an interdisciplinary programme that explicitly considered human beings in relation to the rest of the biosphere. Proposals from the conference developed into a plan for a MAB programme, which was approved by the General Conference of UNESCO in 1970. Soon after its establishment in 1971, the structure of the MAB programme was defined. It was divided into fourteen 'Project Areas' each relating to human interactions in specific zones of the biosphere (e.g. temperate and Mediterranean forests, tropical and subtropical forests, islands, urban systems) or to processes occurring in all parts of the biosphere (e.g. conservation of natural areas and genetic material, pest management and fertilizer use, perception of environmental quality).

More than 20 years after the establishment of the MAB programme, few reviews of the programme or its component Project Areas (referred to below as 'Projects') have been undertaken. On its tenth anniversary, a major symposium reviewed its history and success to date (di Castri et al., 1984) and a General Scientific Advisory Panel was established. This published a formal appraisal of the programme in 1986 (UNESCO, 1986a). In 1991, MAB was again reviewed by ISCU's Scientific Committee on Problems of the Environment (SCOPE, 1992).

One of SCOPE's recommendations was that reviews of a small number of MAB's Projects should be undertaken. The purpose of such reviews should be to evaluate the success of the Projects in scientific and organizational terms, in order to provide information for guiding MAB's future. Such information would also be invaluable in the development of other international interdisciplinary research programmes involving both social and natural sciences, many of which are currently planned or under development (Price, 1990a). The International Co-ordinating Council (ICC) of the MAB programme made a similar recommendation at their twelfth session in 1993 (UNESCO, 1993). A particular reason for the need for a review of individual MAB Projects is that work under the aegis of MAB has been highly decentralized, being directed through National MAB Committees, and that the flow of information to the MAB Secretariat in Paris has been rather uneven. This also applies to the archiving and abstracting of documents and other information deriving from MAB Projects.

With this background, this book provides an initial review of the organization of research and its scientific results within MAB Project 6 (MAB-6), the focus of which is on mountain areas. The review is further limited in its spatial focus, considering only the mountains of nine European countries – Austria, Bulgaria, the Czech Republic, France, Germany, Poland, Slovakia, Spain and Switzerland – and the former USSR. It is in these countries that the majority of MAB activities in European mountain areas have been conducted.

The review is based on the analysis of published and unpublished documents, and interviews and field visits in the ten countries in 1992 and 1993. It is hoped that this regional review will be the first component of a global review of MAB-6 – which has also included activities in Africa, Asia, the Americas and Oceania – and, further, that the bibliography will be of particular value by providing a point of entry to a vast amount of research, generally little-known, which has been conducted over the past two decades in the mountains of Europe and the former USSR.

MAB IN THE MOUNTAINS: BACKGROUND

Project 6 was one of the first MAB Projects to begin its activities. On the recommendation of the first session of the ICC, a 'panel of experts' on MAB-6 met in Salzburg, Austria, for a week in January/February 1973 (UNESCO, 1973a). A draft report from this meeting was presented to the second session of the ICC in April 1973. At this meeting, the ICC decided to transfer tundra ecosystems to Project 6, the title of which thus became 'Impact of human activities on mountain and tundra ecosystems'. The ICC also identified Project 6 as one of four to which immediate attention should be given, by convening a temporary working group to plan the broad outline of an international co-ordinated programme, consider preliminary syntheses and recommend methodologies and plans of study to MAB National Committees wishing to contribute to the programme (UNESCO, 1973b).

The working group met in Lillehammer, Norway, in November 1973 (UNESCO, 1974). This was its only meeting and, although it recommended a 'flexible international task force to ensure continuity and further development of internationally co-ordinated activities' within the framework of MAB-6, this was never constituted. However, regional MAB-6 meetings in western Europe, Nepal, the United States and South America followed in later years, and many participants in early MAB-6 meetings have continued to communicate and co-operate through the International Mountain Society, founded in 1981, and by other means (Ives and Messerli, 1990).

At Lillehammer, the working group identified three 'problem areas' as a basis for the development of a core programme of internationally co-ordinated activities:

(1) Resource development and human settlement in tropical mountain regions;

(2) Tourism, technology and land-use alternatives in temperate mountains; and

(3) Land use in high-latitude mountain and tundra ecosystems, with special reference to grazing, industrial development and recreation.

Clearly, the second problem area is one of primary concern in Europe and the former USSR, though high-latitude mountains also exist in the latter and in Scandinavia. The working group noted that the problem areas were 'not intended to be polished project plans for core programme activities, but rather a number of ideas and propositions for further elaboration' (UNESCO, 1974: 18).

The general objective for work in temperate mountains was 'to give a better knowledge of the mountain ecosystems in the temperate zone in relation to human use, so that present ecosystems can be conserved and new stable ecosystems devised to replace old systems which are no longer socially relevant and economically viable. In essence, this research should make it possible to give management prescriptions to achieve desired aims in the management of the ecosystems of mountains in the temperate zone'. It was proposed that research should concentrate on the problem of the impact of tourism, through the co-operation of natural and social scientists. Particular attention needed to be given to critical factors and threshold points in the process of change, through analysis of carrying capacity. A prime concern was to focus on the needs of local people and their control of the environment or the means by which this control is lost. The main framework was to be the spatially integrated unit: a village and its hinterland, a mountain valley, or a region with similar or contrasting levels of change. Serious attention also needed to be given to the time dimension (UNESCO, 1974: 33–34).

ORGANIZATION OF THE BOOK

Following this introduction, the book is divided into three parts. Part I considers MAB-6 Projects in western Europe, and Part II describes MAB activities in the mountains of eastern and central Europe and the former USSR. As explained in more detail below, MAB-6 activities effectively have not taken place in Bulgaria, the Czech Republic, Poland and Slovakia, so that the description of MAB mountain activities in these countries focuses on mountain biosphere reserves. Thus, the division of the nine chapters which form the core of the book is not based merely on geographical contrasts between different regions of Europe; it also derives from significant differences in the implementation of MAB Projects, as well as the intended co-ordination of MAB-6 Projects in the former socialist countries.

The chapter for each country is divided into four sections. The first briefly describes the location and topography of the mountains, together with recent and current topics of concern at the interface between environment and human activities that is the core of MAB. The second describes the formal organization of MAB activities, particularly with respect to MAB-6. The third provides a brief description of MAB-6 study areas and/or mountain biosphere reserves and the research conducted within these regions. The last section presents conclusions about the results and implications of these MAB activities.

Part III draws the previous chapters together, examining institutional, organizational and scientific issues. It also presents conclusions, not only on how the proposals of the Lillehammer working group were interpreted in Europe and the former USSR, but also with respect to the lessons for interdisciplinary research that may be learned from the diverse activities undertaken within the context of MAB in mountains ranging from the Pyrenees to the Siberian Far East.

PART I

MAB-6 Projects in Western Europe

CHAPTER 2

WESTERN EUROPE: AN INTRODUCTION

The majority of participants at the MAB-6 Salzburg and Lillehammer meetings in 1973 were national experts nominated by MAB National Committees. Many of these representatives were from western European countries, including all of those considered in this report except Spain; and also Finland, Italy, Norway, Sweden and the United Kingdom. At the Lillehammer meeting, the Austrian and French representatives both mentioned projects under development in the Alps. The Norwegian representative noted that, although work within the IBP of ICSU was being undertaken in mountain areas, the focus for Norwegian research within the MAB programme would be in Arctic areas. Both the German and British representatives indicated an interest in undertaking projects within the scope of the programme (UNESCO, 1974).

By 1978, MAB-6 research projects were underway in Austria, France, Italy and Switzerland (Glaser, 1978), and Spain. This year was also marked by two meetings which brought together MAB-6 researchers from different western European countries, in Grindelwald, Switzerland (June), and Berchtesgaden, Germany (November). The seminar in Grindelwald was part of the European Conference of Ministers Responsible for Regional Planning. It considered 'pressures and regional planning problems in mountain regions,' and was sponsored by the Council of Europe (1978). Papers at the Berchtesgaden meeting considered ongoing projects in Austria, France, Italy and Switzerland and the planned project in Germany (German National Committee, 1979).

In 1981, two meetings focusing on MAB-6 in western Europe were held, in Vallouise, France (October) and Berchtesgaden (December). Papers presented at these meetings described Austrian, French, German, Italian, Swiss and also Soviet projects (German National Committee, 1982; Ministère de l'Environnement, 1985). Papers describing Swiss projects were also presented at the tenth anniversary symposium (Chapalay, 1984; Darbellay and Stucki, 1984; P. Messerli, 1984). At the next international MAB-6 meeting, at Pays d'Enhaut, Switzerland, in 1983, presentations were given on the German and Swiss programmes (Messerli and Stucki, 1983). Further European MAB-6 meetings were held in Berchtesgaden in 1984 and 1985. Papers at the 1984 meeting summarized the current status of projects in Austria, France and Germany, as well as recent activities in Bulgaria, Nepal and the USSR (German National Committee, 1985a). The 1985 meeting included papers summarizing the current status of projects in Chile, France, Germany and Switzerland (German National Committee, 1985b).

Berchtesgaden was also the site of the first meeting of 'European' MAB Committees (EuroMAB) – which also include those of North America – following the MAB-ICC's designation of co-operation between European countries as a priority theme at its 1986 meeting (UNESCO, 1986b). Following this first EuroMAB meeting in 1987, they have been held biennially in Třebon, Czechoslovakia (1989), Strasbourg, France (1991), and Zakopane, Poland (1993). The latter was followed by an international symposium on ecological zonation in mountain areas, with specific reference to global environmental change.

Meetings on the subject of biosphere reserves, beginning with the first international biosphere reserve congress in 1983 (UNESCO-UNEP, 1984), have provided other opportunities for communication between scientists working in the mountains of western Europe. Biosphere reserves derive from MAB Project 8 (conservation of natural areas and the genetic material they contain), although the concept has gone significantly beyond the original version which, in a similar manner to MAB-6, was elaborated by an international taskforce in 1974 (Batisse, 1986). Mountain biosphere reserves have been designated in all of the countries considered in this review. Although a considerable amount of research has been conducted in these reserves, not much of this is described in the chapters which follow, as these concentrate specifically on MAB-6 projects.

CHAPTER 3

AUSTRIA

MOUNTAIN REGIONS

The mountains of Austria occupy more than half of the nation's area (Figure 3.1). Only parts of the provinces of Upper and Lower Austria, Burgenland and Styria, which include the largest cities – Graz, Linz and Vienna – are low-relief terrain. The other major cities – Innsbruck, Klagenfurt and Salzburg – are either in or adjacent to mountains. The westernmost mountains are in the Vorarlberg, on the border with Liechtenstein and Switzerland, and comprise the Vorarlberg Alps (highest peak: Rote Wand, 2,704 m) and the Silvretta (Fluchthorn, 3,399 m). To their west are the

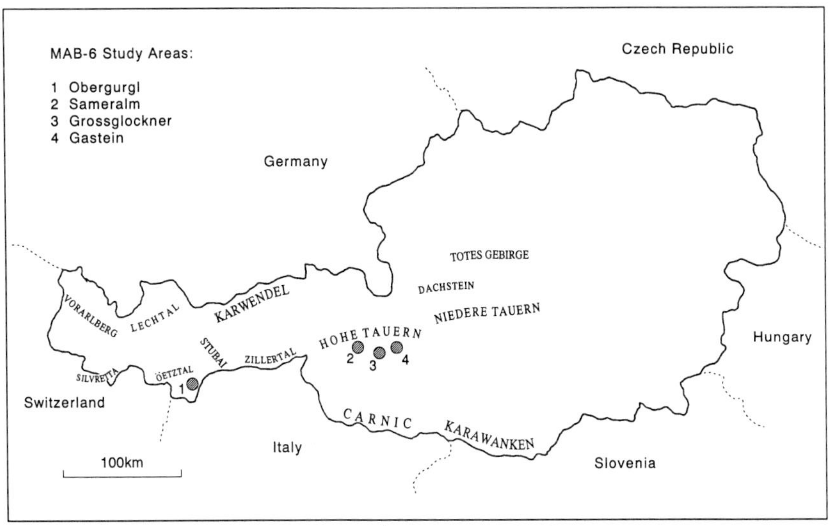

Figure 3.1 Austria: mountain ranges and MAB-6 study areas

mountains of the Tyrol, spanning the country from the German border in the north to the Swiss and Italian borders in the south. The northern groups are the Lechtal Alps (Parseier Spitze, 3,038 m) and the Karwendel, north of Innsbruck (Bettelwurf, 2,725 m). South of these are the Stubai Alps (Zückerhütl, 3,507 m) and the glaciated Oetztal and Zillertal Alps (Wildspitze, 3,774 m; Hochfeiler, 3,510 m) on the Italian border.

The highest mountain in Austria is the Grossglockner (3,797 m) in the glaciated Hohe Tauern, which separate the provinces of Salzburg and Carinthia. Their eastern extension are the lower Niedere Tauern (Hochgollig, 2,863 m) in Styria, to the north of which are the Dachstein (Hohe Dachstein, 2,996 m) and the Totes Gebirge (Grosser Priel, 2,514 m). Finally, along Carinthia's southern margin, are the Carnic Alps (Hohe Warte, 2,780 m) on the Italian border and the Karawanken (Hochstuhl 2,236 m) on the Slovenian border south of Klagenfurt.

The mountains of Austria have considerable areas above timberline, including a number of significant glaciers and glacier complexes. In recent decades, there have been considerable changes in the uses of these mountain landscapes. In particular, tourism has become the dominant economic sector. One unusual aspect of tourism in the Austrian Alps is its dependence on foreign visitors, who account for about three-quarters of the total, of whom two-thirds are from Germany. As in other alpine countries, numbers of tourists increased quite rapidly during the period of growing winter demand until about 1980, and have subsequently remained static or declined. This trend has been linked mainly to the decreasing dominance of the summer season in recent years – especially in Salzburg, Tyrol and Vorarlberg – exacerbated by the reallocation of both government and private resources to winter tourism. Another trend has been an increase in day and weekend tourism, which leads to the overloading of transport and tourist infrastructures. Attempts to cope with peak demands have often caused economic problems because of the high maintenance costs of rarely-used facilities.

In some ways, tourism is quite well-integrated with the agricultural sector, particularly through opportunities for part-time employment and the fact that much accommodation is in the large family houses that are typical of the Austrian Alps. However, younger people are becoming less interested in continuing this traditional pattern, especially as real estate prices rise because of demand for second homes and the market for tourism becomes more competitive, and they are tending to leave the mountains. Nevertheless, private land owners are often recompensed for the use of their land for winter tourism, and subsidized investments in the restoration and improvement of pastures can lead to increases in the potential availability of fodder. These relationships vary considerably between the different provinces, each of which has different sets of regulations, and also between different communities, which have considerable control over local land-use zoning.

Tourism has had many effects on mountain environments. The visible impacts derive from the construction of hotels, second homes, roads and various skiing facilities. These have included glacier-skiing areas, which have been developed since 1966 and permit year-round skiing. Their development has been associated with both visual and physical pollution, although techniques to minimize the latter have been increasingly implemented in recent years in response to environmentalist pressure and regulation. At lower altitudes, the creation of ski-runs has had substantial local effects on vegetation, forests, animals and risks of soil erosion. While initial work on the development of appropriate revegetation and land management approaches in damaged and sensitive areas has begun, it has, as yet, had relatively little impact. Overall, there is increased recognition of the importance of maintaining agriculture as an essential contributor to all resources required by tourism.

The status of forestry varies greatly across the Austrian Alps. There are considerable differences in patterns of ownership; particularly high proportions of forests in the east are federal and, in the west, private (often in very small parcels). Although forestry is generally declining, the provinces have funded increasing road construction in recent years, linked especially to the need to remove trees affected by acid precipitation. However, while the general intention is to remove only affected trees, the opportunity afforded by the new roads has often been used to expand clear-cutting. At present, timber prices are very low, and owners are becoming increasingly interested in using financial assistance, e.g. from government subsidies or taxes deriving from tourism, to set aside designated areas of forest for conservation purposes.

ORGANIZATION OF MAB-6 ACTIVITIES

The National MAB Committee was established in 1972 by the Austrian Academy of Sciences, which appoints its chairman. There have been four chairmen to date, each from a different scientific discipline. The members of the committee now include both individuals designated by federal ministries and provinces interested in current activities, and scientists appointed by the chairman, mainly on the recommendation of existing members. This pattern has existed since 1976; previously, the committee was drawn from the members of the Academy and there were no provincial government delegates. The proportion of provincial government delegates increased again in 1983, since when they have been in the majority. In recent years, the proportion of scientists has been increasing again. There is no specific secretariat for the MAB programme; necessary work has always been done by Dr Gerlinde Schrammel of the Academy of Sciences, who has been responsible for all international scientific programmes since the early 1970s.

The National Committee usually meets twice a year. In order to maximize the value of the limited funds available from the federal government (mainly the Ministry of Science and Research), it defines topics of interest for four- to six-year periods. The first of these topics was mountain ecosystems, for which a member of the Academy was appointed as co-ordinator in 1973. This topic remained important until 1983, when political imperatives led to a change in emphasis, initially to the effects of hydroelectric generating facilities. Subsequent topics of interest have been limnological resources, ecotones and, most recently, forests.

Scientists are annually invited to submit proposals within the scope of the current topics of interest. In the 1970s, when most of the MAB-6 research was done, proposals were only subject to internal review. They now go through an international review process. Recommendations on proposals are considered by the National Committee when it decides on the distribution of funds. It is important to mention that, although proposals can be for up to 4 years, funding has always been for 1 year at a time; a new proposal is needed every year even for ongoing projects. It is also worth noting that the province in which a research project is to be conducted has to approve of the proposed work. All MAB research has been funded by the Federal Ministry for Science and Research, with some contributions from the Federal Ministry for Health and Environmental Protection, the Fund for Scientific Research and, in some cases mentioned below, provinces and other institutions. All funding has been directed through the National Committee.

MAB-6 RESEARCH ACTIVITIES

MAB-6 research programmes in Austria have been conducted in three Alpine regions: the valley of Obergurgl in the Oetztal Alps (Tyrol); the Grossglockner and Gastein areas of the Hohe Tauern in Carinthia and Salzburg, respectively; and the Sameralm area of the Salzburg Limestone Alps (*Kalkalpen*) (see Figure 3.1). The research in each of these areas is described below. In addition, a few studies of human physiological responses to activity at medium and high altitudes were undertaken by scientists from the Institute of Zoology at the University of Salzburg in the early 1980s. These, together with a study of the demography of the Hohe Tauern from 1900 to 1971 undertaken by the Institute of Geography of the University of Vienna, are summarized in Franz (1985). This is one of the series of

Figure 3.2 Obergurgl in 1977 (photograph by Werner Lohmann)

publications of the Austrian MAB programme (*Veröffentlichungen des Oesterreichischen MAB-Programmes*), published by Universitätsverlag Wagner in Innsbruck. Together with a number of interviews (see Author's Preface and Acknowledgements), these reports are the principal source for the descriptions of projects below.

Obergurgl

Obergurgl is a village in the Oetztal Alps of Tyrol, at an altitude of about 2,000 m (just below timberline) at the head of the Upper Oetztal, with peaks to 3,700 m (Figure 3.2). The pattern of life probably changed relatively little from the twelfth century, when the valley was settled, into the mid-twentieth century. The economy was a traditional one, based on agriculture and transhumant grazing, and the population of about 120 changed little in size. In the early 1950s, Obergurgl was one of the many villages where winter tourism took hold. Its rapid growth was linked to the construction of hotels, ski-lifts, and many ski-trails and the decline of agriculture.

By the 1970s, only six out of eighty families continued to farm. The resident population was about 300, with a seasonal population of 800 migrant workers. Hotels and guest-houses provided 2,000 beds for 40,000 visitors a year, who mainly came in the winter season, which accounted for 80 per cent of revenues. Much of the former agriculturally productive land that was not avalanche-prone had been built on. Many alpine meadows and pastures had been destroyed; others were experiencing decreasing productivity and erosion, primarily due to skiing, but also because of summer hiking. These trends were close to four thresholds (Moser and Moser, 1986):

(1) Substantial expansion of accommodation would have required much greater building densities in already built-up areas;

(2) The expansion of skiing would have required much higher investments than before, because of the existing density of development;

(3) Pressures on farming were close to leading to its demise, thus threatening the maintenance of a 'typical' alpine landscape; and

(4) The social fabric was close to breaking down.

In the 1960s, the University of Innsbruck had constructed its Alpine Research Station in Obergurgl, notably for work within ICSU's IBP. The IBP project concentrated on the Hoher Nebelkogel (3,184m) and led to the collection of a considerable volume of scientific data concerning this mountain. The availability of these data, combined with the increasing anthropogenic pressures on the environment and society of the valley, motivated the initiation of a new project in 1971. Its co-ordinator was a botanist who had been born in the Oetztal. The project's objectives were 'to study in an integrated fashion the daily life and future options of an alpine village which had become a tourism centre, and to use the systems knowledge thus gained to raise awareness and provide a better basis for individual and communal decision-making in the best long-term interest of the villagers' (Moser and Moser, 1986: 102). In 1973, the project finally became a MAB-6 project, although the overall co-ordinator for mountain programmes in the National Committee did not favour it greatly.

The methodology chosen for the project was systems analysis, which had already been applied to the ecosystems of the Hoher Nebelkogel during the IBP project. In 1974, the National MAB Committee and the International Institute for Applied Systems Analysis (IIASA: near Vienna) sponsored a 5-day workshop to 'develop a preliminary model of human impact on a simple alpine ecosystem and the policy options, by combining the knowledge and insights of business men, government officials, and scientists' (Himamowa, 1974: 2–3). The participants included people from Obergurgl, representatives of local government, scientists from Austria and other European countries and Canadian systems analysts from IIASA. The model developed during the workshop included four major components: recreational demand; population and economic development; farming and ecological change; and land-use and development (Figure 3.3).

The model led to the identification of problem areas and consensus on eight ranked research priorities (Himamowa, 1974: 44–45):

(1) Sociology of villagers in relation to attitudes about land ownership, emigration and economic opportunities;

(2) Perception of environmental quality by visitors and tourists, initially through photographic scenarios;

(3) Basic mapping of ecological conditions, especially in relation to ski development and soil erosion;

(4) Determination of primary production of pastures and alpine meadows in relation to grazing by wild and domestic animals;

(5) Projection of potential recreational demands in relation to changing transportation systems and public attitudes across Europe;

(6) Continued policy analysis of alternative development schemes and research priorities;

(7) Experimental studies involving manipulation of grazing patterns, trampling of meadows by people and construction activities; and

(8) Economic analysis of the village in terms of employment, savings patterns and costs of hotel construction.

The eighty scientists involved in the project were divided into three groups (Moser and Peterson, 1981):

(1) Co-ordination: administration and logistics, integration and distribution of data flow, co-operation with local people;

(2) Natural scientific studies: effects of summer tourists and skiers on ecosystems (botany, entomology, zoology, microbiology, pedology, meteorology); and

(3) Social scientific studies: ethnology, anthropology, geography, sociology, regional planning, economic theory.

The project was directed by a commission including the co-ordinators, local people and representatives of the Tyrol government, who met five times a year and allocated funds according to the progress of the research. Funding for the study came not only from the federal government, but also from the Province of Tyrol.

Unfortunately, the field component of the project was prematurely

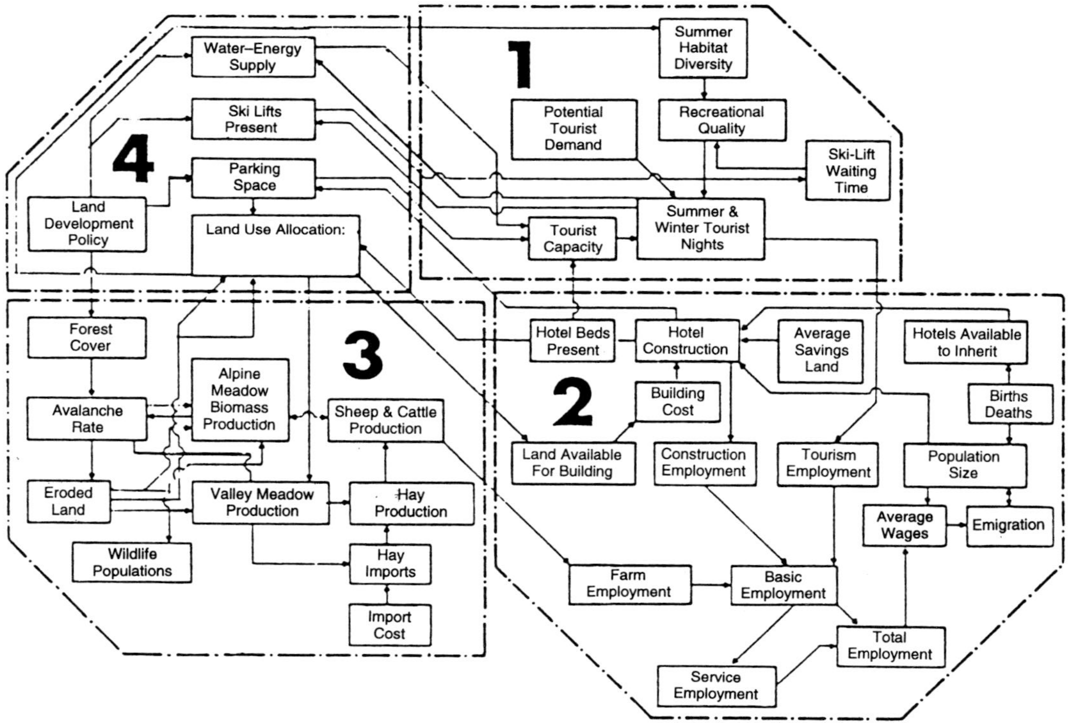

Figure 3.3 The Obergurgl model. The four components are (1) recreational demand, (2) population and economic development, (3) farming and ecological change, and (4) land use and development (from Himamowa, 1974, with kind permission from IIASA)

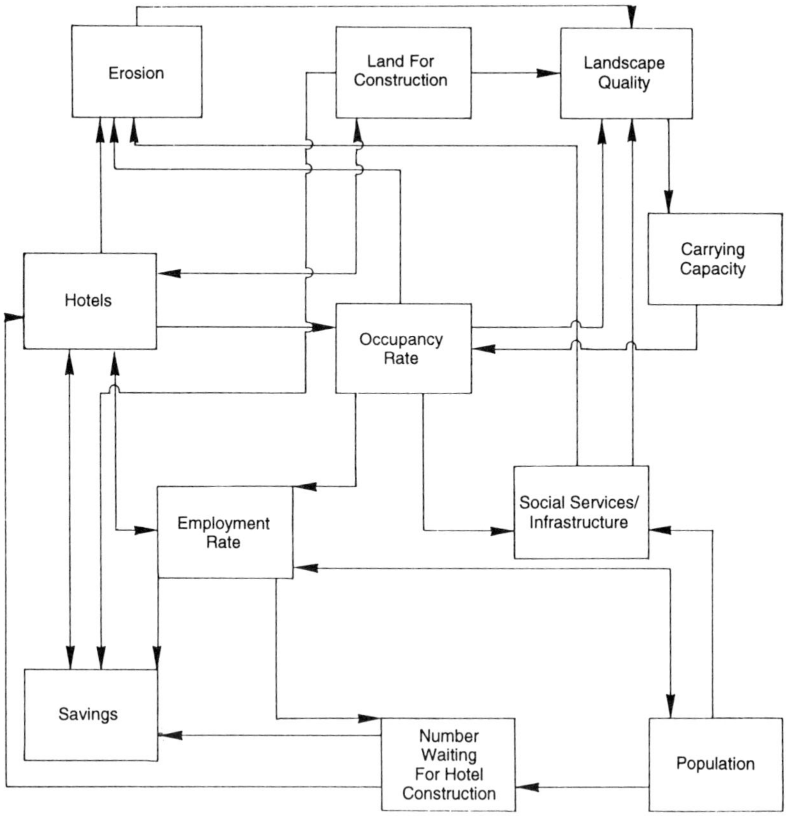

Figure 3.4 The first revision of the Obergurgl model (redrawn from Moser, 1987, with permission)

brought to an end in summer 1979, just after a meeting of international journalists, supported by UNESCO, in Obergurgl. At this point, most of the intended natural science research had been completed, but much of the social science research was in its initial stages. There were a number of reasons for the termination of the project, including internal political problems and the unwillingness of the National MAB Committee to guarantee the financial support necessary to bring the project to its conclusion. These led to the final reason: the project co-ordinator's decision to move to Canada. His successor was told that funding would not be continued, and to close down the project within a year.

The final report of the project brings together the results of the research until 1980. It was published in 1987, and includes fifteen chapters with short English summaries (Patzelt, 1987). The first of these is a brief history of the project, which mentions that the original model was modified twice (Moser, 1987). The first revision, in 1975 (Figure 3.4), was to simplify the model for the integration of field data and for discussion with local people. The revised model was discussed at a second international workshop, at which various scenarios were presented and discussed. Many other workshops, involving scientists and policy-makers from various European countries and New Zealand, followed. A second revision was made in 1978, and was discussed at an internal project workshop. The intention was to test and revise further this model using the quantitative data deriving from the project research activities and compiled in a database. However, this final stage did not occur because of the cessation of funding, and the database was closed down. Much of the data it contained were never analysed or published.

The remaining chapters of the final report cover a wide range of topics. Those deriving from the natural scientific studies consider: energy balances and budgets during the growing season; soils; vegetation (mapping and primary production); and the impacts of tourism (skiing and trampling) on alpine vegetation, soils, invertebrates and grouse. The few papers from the social scientific studies consider socio-economic changes (demography, employment, tourism); changes in value systems relating to the development of tourism; physical anthropology; and land use and hunting.

In spite of the project's premature demise, it has had a number of significant results. The first of these was the development of an integrated model-based approach to the analysis and understanding of alpine environments, involving local people and decision-makers as well as scientists. This links to a second result: recognition of the need for direct communication between scientists undertaking research in an area and those living in, and responsible for, it. Such communication has continued since the MAB Project, throughout the whole Oetztal. A third result has been the utilization of the findings of the research in developing innovative planning approaches, such as the Tyrolean government's designation of a *Ruhegebiet* (quiet area) in the Upper Oetztal, where traditional agricultural practices are permitted, but the construction of facilities for tourism is not. Finally, as noted by Moser (1987), the project gave the local people a greater awareness of the relative benefits and costs of tourism, changed investment patterns and led to the creation and renewal of many communal activities. It is not clear whether this heightened awareness of the implications of development continues: a new phase of hotel construction began in the late 1980s (Ives and Messerli, 1990).

Grossglockner

The Grossglockner area of the Hohe Tauern National Park is traversed by a scenic road that crosses the highest pass in Austria, currently used by about

Figure 3.5 The MAB-6 Wallackhaus study area (2,300 m) adjacent to the Grossglockner Hochalpenstrasse, looking south (photograph by Christian Körner)

1.3 million people in summer (Figure 3.5). The road also provides easy access to the area for field research, and there are accommodation facilities on the south side of the pass, in Carinthia. Field research has been undertaken in this area at least since the 1930s. Funding for MAB-6 research came not only from the federal government, but also from the Provinces of Carinthia and Tyrol, and the Research Institute Gastein-Tauern Region. The MAB-6 Project in the Grossglockner area included two sub-projects. One was process-oriented, focusing on a number of test sites, with comparative studies in the Gastein area (described in the next section), while the other was more analytical and descriptive. The research involved scientists from Innsbruck, Munich, Salzburg and Vienna.

The process-oriented sub-project built on the preceding IBP projects on the Hoher Nebelkogel (above Obergurgl) and the Patscherkofel (above Innsbruck), and was co-ordinated and undertaken by a group established at the Institute of Botany of the University of Innsbruck in 1973. The work was divided into a synoptic phase, followed by a detailed research phase. The former, starting in 1974, used relatively simple methods, essentially comparable to those used in the IBP projects, to provide an overall description of the biophysical environment. Building on these studies, the research phase (1974–76) began with the development and field testing of a range of new methodologies. This involved intensive ecosystem studies in specific locations, with the intention of developing the best possible understanding of the flora, fauna and micro-organisms of alpine grassland ecosystems, particularly for application in the management of the National Park. The first major field season was 1976, when comparative studies of stands along environmental gradients in alpine pastures were conducted on the alpine grass heath at Wallackhaus (2,300 m). The resulting report includes a number of studies of macro- and microclimatology, botany, plan physiology and productivity, pedology and entomology (Cernusca, 1977). The work continued in 1977 and 1978, but did not evolve into a planned experimental phase because the MAB committee did not approve the necessary funds and the focus of research was moved to the Gastein area, as described below.

A second group of studies, building on the work begun in the 1970s, considered an altitudinal transect with six sites, at altitudes from 1,085 to 2,528 m, on the south side of the Grossglockner pass. The research took place from 1978 to 1983, when funding ceased. It was undertaken by scientists from the full range of natural-science field disciplines from the University of Innsbruck, co-ordinated through the Institute of Botany, as well as the Munich-Weihenstephan Technical University and the Bavarian Water Authority. This research, together with continued work at Wallackhaus, is described in Cernusca (1989a), which includes an English overview (Cernusca, 1989b).

The primary theme of research was the structure and function of grassland ecosystems at different altitudes. This involved research on water and energy budgets (e.g. Figure 3.6); phytomass, canopy structure (e.g. Figure 3.7) and microclimate; energy content of plant material; fodder value of plant canopies; carbon budgets; canopy and soil respiration; soil microbiology; and invertebrate zoology. A second theme considered anthropogenic stresses on grassland ecosystems: grazing by cattle and sheep; abandonment of pastures; skiing and snow-making; hiking; and pollution from vehicles and road maintenance. A third theme was the application of research from the other themes for the management of the Hohe Tauern National Park.

The first series of studies in the other sub-project in the Grossglockner area was undertaken from 1974 to 1978. The co-ordinator was the overall co-ordinator for MAB-6 in Austria, from the University for Bodenkultur in Vienna. Other participants in the study came from this institution, the Austrian Society for Nuclear Studies and the Federal Forestry Research Institute in Vienna, and the Institute for Botany at the University of

Figure 3.6 (Opposite) Typical diurnal courses of evapotranspiration (plant communities) and evaporation (pan and bare soil) on bright midsummer days at various altitudes in the Grossglockner area (from Körner et al., 1989, with permission)

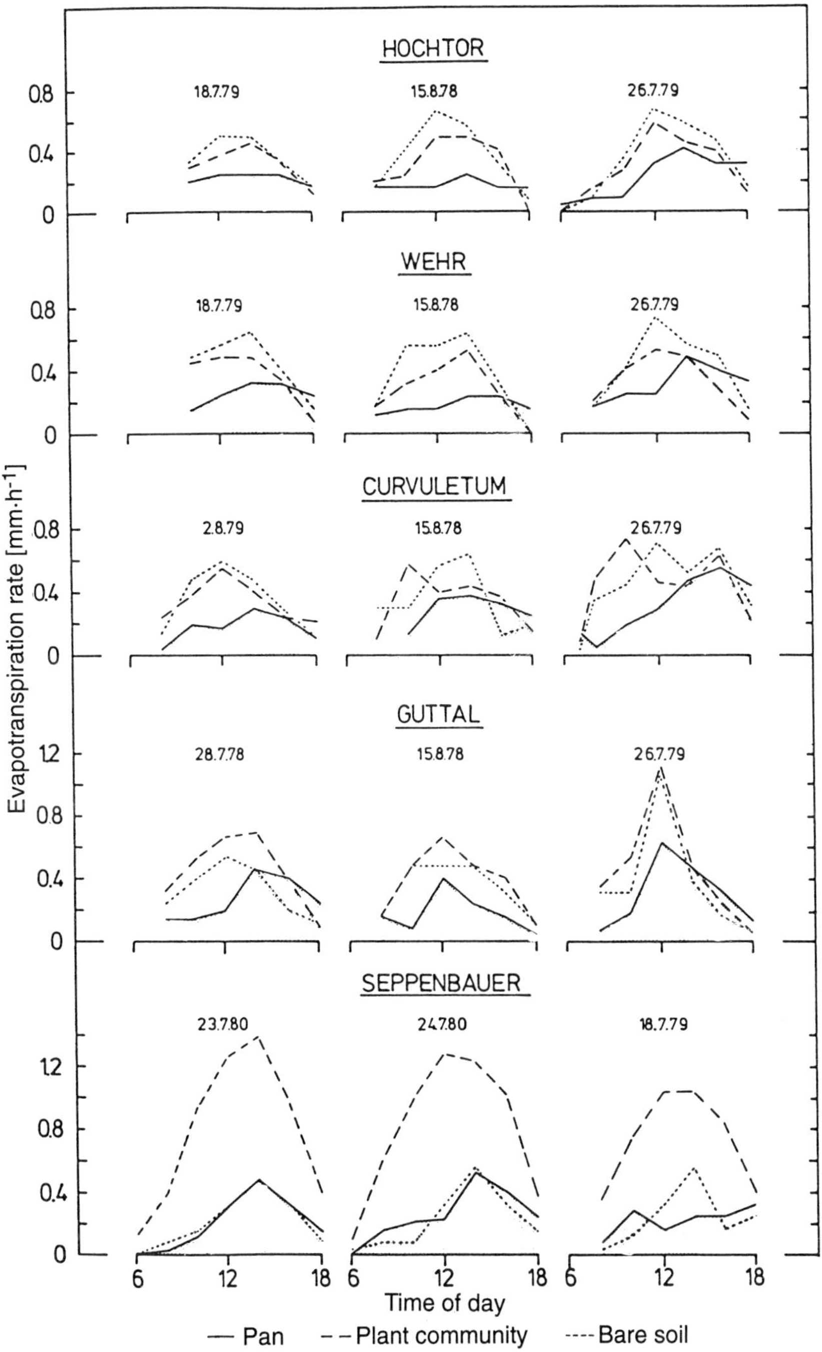
HOCHTOR
18.7.79
15.8.78
26.7.79
WEHR
18.7.79
15.8.78
26.7.79
CURVULETUM
2.8.79
15.8.78
26.7.79
GUTTAL
28.7.78
15.8.78
26.7.79
SEPPENBAUER
23.7.80
24.7.80
18.7.79
Evapotranspiration rate [mm·h-1]
0
0.4
0.8
1.2
6
12
18
Time of day
Pan
Plant community
Bare soil

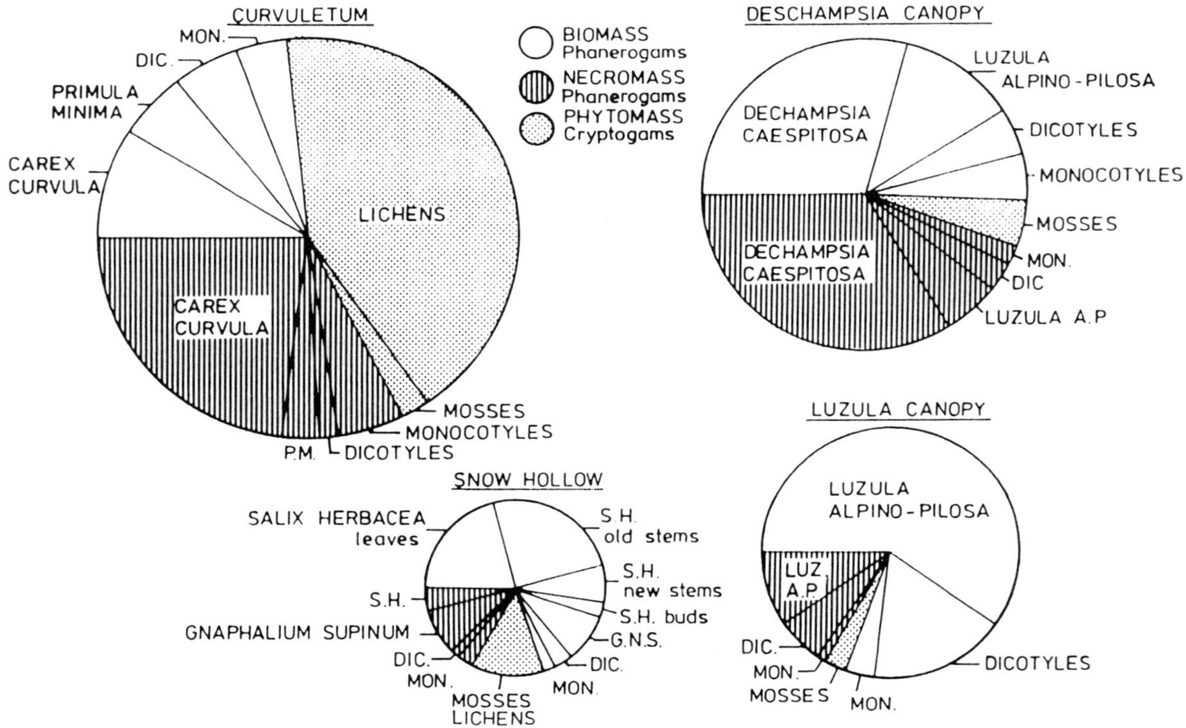

Figure 3.7 Above-ground phytomass of the plant communities of the Grossglockner area and its distribution among different components. The area of the circles is proportional to total above-ground biomass (from Cernusca, 1991, with permission) Dic., Dicotyles; Mon., Monocotyles; P.M., *Primula minima*; Luzula A.P., *Luzula alpino-pilosa*; S.H., *Salix herbacea*; G.N.S., *Gnaphalium supinum*

Innsbruck. This work resulted in two reports describing studies at various scales, including both regional-level research and studies with quite limited spatial scales, particularly at the Wallackhaus (2,300–2,500 m) and the forefield of the Pasterze glacier (2,000–2,400 m). The first consists of fourteen papers describing the overall environment (climatology, soils, vegetation); soil chemistry and its dynamics; and the water economy of meadows at and above timberline (Franz, 1980). The second includes fourteen papers on soil zoology and the ecology of soil micro-organisms (Franz, 1981).

Other products of the first series of regional studies were a report on the climatology of the region, based particularly on research from 1974 to 1980 (Dobesch, 1983): and a report including maps at scales of 1 : 5,000 (soils, topography), 1 : 25,000 (vegetation) with explanatory text, together with discussions of the use of aerial photography and LANDSAT images (Austrian MAB programme, 1983).

Analytical/descriptive research, by scientists from various institutes – including the Zoological Institute of the University of Salzburg, the Institute for Plant Ecology and Sociology at the University for Bodenkultur in Vienna, and the Institute for Phytosociology in Klagenfurt – continued in the Grossglockner region until 1981. This research included studies summarized in Franz (1985), on small birds (community structure, territoriality and human impacts); vegetation complexes and primary productivity of alpine pastures; and the impacts of heavy metal and air pollution on vegetation and lichens along the Grossglockner road.

Since the conclusion of the field research described in the reports mentioned above, work has continued on the integration of its results, particularly maps, using geographic information systems (GIS) (Beissmann, 1989). The MAB-6 maps have been integrated with remote-sensing imagery from satellites and aerial photographs, as well as maps from the 1930s to 1950s, in order to develop an historical perspective. A particular emphasis has been the assessment of the consistency of classification of different thematic maps and of possibilities of combining them. The results show the difficulties of combination in a topographically complex area, especially when the maps were not prepared using orthophotos or global positioning systems (the latter had not been developed when the MAB-6 research was done). The results also provide bases for optimizing the location of future field-work and show the possibilities of developing interdisciplinary case- and rule-based expert systems for defining relationships between the occurrence of different components of alpine ecosystems. It is possible that this synthesis of some of the results of the MAB-6 work will be published by the National MAB Committee in the near future.

Figure 3.8 Part of Schlossalm ski area, Gastein MAB study area (photograph by John Haslett)

Gastein

The thermal mineral waters of the Gastein valley on the north side of the Hohe Tauern were first developed in 1350. Spa-based tourism began in 1830 and has continued ever since. Subsequently, the valley has also been developed for both summer and, since 1912, winter tourism. Summer tourism was dominant until the late 1960s, but has subsequently stagnated, and winter tourism has become more important in this major ski resort. The development of tourism has been linked to the decline of agriculture and damage to the forests. One reason for the choice of the Gastein valley for MAB-6 research was the extent of ecological impacts deriving from skiing, in contrast with the relatively unchanged Grossglockner area. Another reason was the existence of the Research Institute of the Gastein-Tauern Region, founded in 1936. The MAB-6 Project marked a partial change in the emphasis of research at this station, from physiological and medical to ecological studies. MAB-6 research in the Gastein area was supported by this station and the government of the Province of Salzburg, as well as the usual federal agencies. The availability of local funding was one of the major reasons for the change in the focus of research from the Grossglockner area to Gastein.

MAB-6 Projects in the Gastein valley took place on three test sites near the timberline (1,800 m): the Stubnerkogel and Schlossalm ski areas (Figure 3.8), and the Nassfeld area, where summer grazing continues. The first project was conducted in 1977 by the same group that had previously done process-orientated research in the Grossglockner area: scientists from the Institute of Botany at the University of Innsbruck (which co-ordinated the work), the Munich-Weihenstephan Technical University and the Bavarian Water Authority. This work considered alpine pastures that had been abandoned, particularly due to the growth of tourism. Most of the work was done on two meadows: one still under management and one that had not been grazed intensively for 13 years, and had been mown once 4 years previously. The resulting report includes twenty-six papers on climate, hydrology, botany (emphasizing primary production and vegetation composition), zoology, soils, microbiology and landscape changes (particularly in relation to the development of ski-slopes) (Cernusca, 1978). This work was continued from 1982 to 1985, focusing on the impacts of downhill skiing on alpine ecosystems (e.g. Figure 3.9). The research was partly designed to complement the work in the Grossglockner area in similar, but relatively undamaged ecosystems (Cernusca, 1989a). The results have contributed to guidelines for the management and regulation of, and financial compensation for, the development of ski areas and trails in Austria.

As in the Grossglockner area, analytical/descriptive studies were undertaken in the Gastein area. These studies, from 1977 to 1981, involved scientists from the Zoological Institute of the University of Salzburg, the Research Institute Gastein-Tauern region, and the Silvicultural Institute of the University for Bodenkultur in Vienna. This work is summarized in Franz (1985), which includes chapters on the hydrology of a small alpine catchment; soil protozoa and nematodes; vegetation mapping; and changes in forest cover based on palynological research.

Later research considered the impacts of downhill skiing developments on forest ecosystems (Mayer, 1990). This work began wih a preliminary study of the silvicultural effects of competitive ski-runs in Zell am See (also in the Hohe Tauern) from 1981 to 1985. The work in the Gastein valley was undertaken from 1984 to 1986 by scientists from the Silvicultural Institute at the University for Bodenkultur in Vienna. The final report includes ten chapters, beginning with a literature review and a general description of the biophysical environment and development of tourism in the region. Subsequent chapters give a detailed analysis of the ecology and hydrology of the ski-runs and surrounding forests, and of damage due to skiing, ski area expansion (runs and lifts) and animals. The final chapters present a methodology for assessing the environmental impacts of ski-runs and its application to the projected expansion of skiing in the Gastein valley.

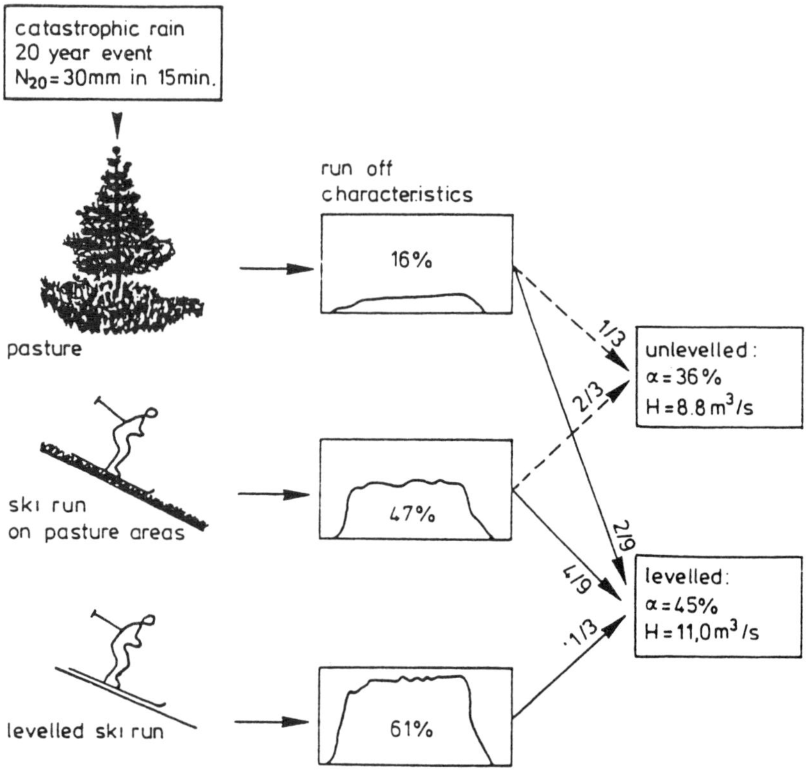

Figure 3.9 Impacts of the Schlossalm ski-area on streamflow characteristics, showing runoff characteristics associated with a 20-year precipitation event. The ski-runs cause an increase in the runoff coefficient (α) and the peak discharge (H). The fractions associated with the arrows designate the percentage of the area covered by the respective vegetation units (from Cernusca, 1991, with permission)

Sameralm

The Sameralm region (980–1,702 m) of the Salzburg Kalkalpen underwent a major socio-economic change from the 1950s, with conversion from an agricultural to a tourism-based economy. This was linked to a change from intensive dairy farming to the extensive breeding of young animals and also the abandonment of pastures which rapidly reverted to forest. These two trends both led to increasing soil erosion (Figure 3.10). The first series of MAB-6 studies in the region took place from 1973 to 1980. The research was undertaken by scientists from the Institute of Geography at the University of Salzburg, which has a field station in the area. The final synthesis report (Riedl, 1983) includes ten chapters describing research results: geology and geomorphology (including hydrology); climatology;

Figure 3.10 Gully erosion maintained by runoff concentrated along disused paths to the alpine pastures in the Sameralm MAB-6 study area (photograph by H. Riedl)

vegetation; evaluation of the natural potential of major sub-areas; cultural geography; recent socio-economic structural change; recent structural changes in pastoral agriculture – and in regional context; the impacts of socio-economic change on the morphology and soils of pastures; and brief conclusions for economy and landscape management.

A second series of MAB-6 'ecogeographical comparative studies' took place from 1981 to 1983, and involved scientists from the Institute of Geography at the University of Salzburg (which co-ordinated the project) and also the Institute of Geography of the University of Vienna and the Institute for Regional Planning and Agriculture of the University for Bodenkultur, Vienna. The final report (Riedl, 1987) includes an introduction and eleven chapters summarizing research on soils; relative humidity; the structure of alpine farming; impacts of tourism on subalpine landscapes; changes in vegetation and soil erosion linked to the expansion of extensive grazing and inadequate pasture maintenance; changes in timberline; and karst caves.

CONCLUSIONS

The MAB-6 projects in the Austrian Alps are notable for the range of approaches applied in the various research areas; whether one considers methodologies, the extent to which individual studies have been undertaken within a defined framework, or the balance between natural and social scientific components. Some of the projects have been of considerable significance both within Austria and at the international scale.

While the initial Sameralm project provided a somewhat integrated approach to a region whose economy had undergone considerable change, the analyses were generally rather qualitative and descriptive. However, the project provides a good overview of the course and results of the change from an agricultural to a tourist economy. The later studies expanded on this overview, but without a research framework. Similarly, the analytical/ descriptive studies undertaken in the Grossglockner area from 1974 to 1981, and in the Gastein area from 1977 to 1981, are essentially a collection of individual studies within a wide range of natural science disciplines. The human component is only included to the extent that the impacts of hiking, grazing and vehicle pollution are considered. Similarly, while the final project in the Gastein area was a coherent analysis of the impacts of skiing on forest ecosystems – and can therefore be described as an integrated analysis with strong (and specified) management implications – human beings were primarily treated as agents of change; the project did not include any social scientific research.

The other two sets of studies both followed on from work conducted within the IBP by scientists from the Institute of Botany at the University of Innsbruck, but developed in very different ways. Like most of the other projects, the process-orientated research which began in 1973 and continued until 1983 in the Grossglockner and Gastein regions considered human beings only as agents of change and did not involve social scientists. These projects provided significant detailed information of relevance to their goal of understanding the functioning of alpine grassland ecosystems, and were a major impetus to the development of ecosystem research in Austria and elsewhere in the Alps. However, the considerable potential for using the results of the Grossglockner research, in particular as baseline data for experimental studies, was unfortunately never realized. Furthermore, these process-orientated studies were multidisciplinary, rather than interdisciplinary; none of them was designed as a coherent whole, and most of the considerable volumes of data that resulted (and were stored in a database) were never linked in quantitative models. The exception concerns models of the effects of canopy structure, which were later further tested in the Caucasus of Georgia (Tappeiner and Cernusca, 1991). In addition, the results of studies on the ecological effects of skiing have been important in the development of regulatory approaches and requirements for financial compensation in regard to the development of the infrastructure of downhill skiing.

The Obergurgl project was unique in many ways. It was already in existence when the MAB programme began, and was included in the Austrian MAB-6 programme although it was not fully supported by the National MAB Committee. It is particularly notable as the first attempt to apply systems modelling for decision-making for a complex mountain system. Not only did the model include people and other biotic and abiotic components; local people and decision-makers as well as scientists were involved in its development. However, many project scientists were not involved in this, and it is unclear to what extent they understood the model or considered needs for providing data for it when designing and undertaking research. The unfortunate early end of the project was due to the interaction of various factors, mostly political in nature, and meant that the final version of the Obergurgl model, or even a sub-model, was not tested with the quantitative data collected during the preceding IBP project, the first seven field seasons, or the process-orientated research in comparable ecosystems in the Grossglockner area. Given the state of development of computers at the time, one might question whether such a model would have been viable; but this question will remain hypothetical. Nevertheless, the international attention given to the Obergurgl model has had many valuable consequences, not least in the later Swiss and German MAB-6 projects.

The Obergurgl project is the only Austrian MAB-6 project that was interdisciplinary and included a substantial component of social science research. To some extent, this reflected the objectives and background of the co-ordinator. There appear to have been at least three linked reasons for the lack of interdisciplinary approaches involving natural and social scientists in other projects. Perhaps the most important was the fact that funding has always been on 1-year cycles. Given the complex process of developing interdisciplinary research, this has been a significant restriction on the evolution of such research in Austria. A second was the limited financial support for MAB projects: in the late 1970s, many scientists, working in four areas, were competing for rather limited funds. Finally, project co-ordinators rarely seem to have been particularly interested in developing integrated programmes, and tended to involve scientists mainly from their own universities – generally from their own institutes.

At the present time, a new MAB pilot project is starting in Austria. It focuses on the degree of naturalness (*Hemorobie*) of forest ecosystems, which cover a considerable proportion of the country's mountains. The pilot phase of the project began at the Institute for Phytophysiology at the University of Vienna in 1992 (Grabherr et al., 1992). If the project is accepted by the National MAB Committee, it is anticipated that its duration will be 4 years (1993–96). The initial objectives of the project would be to define the degree of naturalness of all of Austria's forests objectively, using existing survey data and field research, combined through the use of geographic information systems. Using scientific analysis and consultation, the resulting information would then be used to develop policy and management (including non-management) approaches in order to maximize the natural characteristics of Austria's forests over the long term. The project would involve decision-makers and scientists from a wide range of disciplines. It is particularly timely, given the effects of acid precipitation on the forests and the socio-economic changes that relate to the evolving economies and changing wood markets in Austria's mountains and are likely to derive from Austria's future membership of the European Union.

CHAPTER 4

SWITZERLAND

MOUNTAIN REGIONS

On Switzerland's north-west border with France are the Jura mountains (highest peak: Mt. Tendre, 1,679 m). To their south and east is the flat Mittelland, on which all of the cities are situated and where about 60 per cent of the population lives. The remainder of the country, three-fifths of its area, is occupied by the Alpine chain, which has less than a quarter of the population (Figure 4.1). Furthest to the west is the Mont Blanc massif,

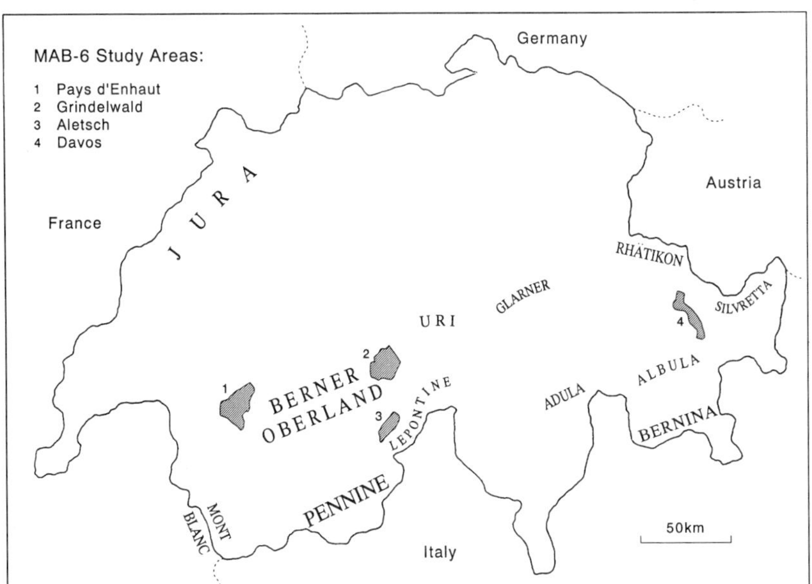

Figure 4.1 Switzerland: mountain ranges and the MAB-6 test areas

whose highest peak on the Swiss/French frontier is the Aiguille d'Argentière (3,896 m). To the east of this massif, on the south side of the Rhône valley along the Italian frontier, are the Pennine Alps (highest peak: Dufourspitze, 4,633 m). North of the Rhône valley are the Bernese Alps or Berner Oberland (Finsteraarhorn, 4,274 m). These two ranges contain most of the glaciers in Switzerland, including the longest in western Europe, the Aletsch Glacier (25 km).

East of the Pennine Alps, on the Italian border, are the Lepontine Alps (Monte Leone, 3,553 m), curving north towards the Uri Alps (Dammastock, 3,630 m) and Glarner Alps (Todi, 3,623 m) and east into the Adula Alps (Rheinwaldhorn, 3,402 m). The remaining Alps are mainly in the Canton of Grisons (Graubünden). From north to south, these include the Rhätikon Alps (Schesaplana, 2,964 m) and Silvretta Alps (Fluchthorn, 3,399 m) on the Austrian border; and the Albula Alps (Piz Kesch, 3,418 m) and Bernina Alps (Piz Bernina, 4,049 m), reaching to the Italian border.

For centuries, human activities have significantly influenced the landscapes of the Swiss mountains; particularly through agriculture and the associated removal of trees for grazing, local use and profit. Today's timberline is 200–300 m lower than its natural limit; the high alpine pastures that are an integral part of the landscape which tourists find so attractive are largely anthropogenic. Since the 1950s, tourism and related activities have become the most important economic sector in the Swiss Alps, accounting for at least 80 per cent of employment in some communities. The dynamism of tourism is one reason that the population of the Alps is growing. However, not all settlements are characterized by this overall positive trend.

The great increase in demand for tourism in the 1960s and 1970s has levelled off in recent years. Nevertheless, the infrastructure, including large numbers of second homes, and many ski resorts, remains. The Swiss Alps' location in the centre of Europe, the excellent transportation network and the diversity and high quality of tourism opportunities mean that tourism will remain dominant; even if there are variations from year to year, depending on both economic and other factors, such as changes in snowfall patterns, which have had considerable effects on the economic base of many ski resorts in recent years. Consequently, discussion of all other socio-economic and resource management issues must inevitably consider their relationships to tourism.

Agriculture, which is heavily subsidized, has been strongly influenced by tourism with respect to: competition for land, particularly in valleys; increases in land prices and rents; and both local and external labour markets. Throughout the Alps, full-time operations have tended to become fewer but larger in recent decades, although numbers of part-time farmers have increased in many areas. There is a widespread trend toward livestock-based intensive agriculture, with conversion of arable land to meadows. Conversely, higher, inaccessible and less-productive meadows and pastures are increasingly being abandoned. The lack of maintenance and the abandonment of fields and pastures have many side-effects, such as increases in the risk of natural hazards and decreases in biodiversity.

The use and management of forests have also been greatly influenced by tourism, for which they are important not only as part of the landscape, but also for the production of raw materials and protection from natural hazards. The latter role has grown greatly in importance in recent decades with the continued expansion of transport, accommodation, energy and other types of infrastructure. Yet, over the same period, the necessary stability of the forests has been endangered by air pollution and decreasing levels of management and harvesting. These derive from the uncompetitiveness of Alpine wood in the international market, substitution of other sources of fuel and construction materials and decreases in the forestry labour force. However, positive trends are beginning to emerge, particularly as a result of the new Forest Law, which subsidizes forestry management activities that are in the public interest.

Another set of major modifications of the Alpine landscape derives from the development of its hydrolectric potential. Nearly all possible dam sites have been used, to provide power not only for mountain communities and the tourist infrastructure, but also for lowland settlements and industries. The hydrological regime has been greatly changed and there are many conflicts beween demands for power, drinking water and sewage treatment, particularly in winter. Other major environmental issues concern pollution from industries both in and around the Alps and also from traffic; a major concern in many mountain resorts in both summer and winter and also along major roads. All of the anthropogenic influences on the environments of the Swiss Alps mentioned above are the subject of diverse sets of policies developed at local, cantonal and federal levels. The increasing recognition of the need to harmonize these policies is gradually leading to results, but this is a complex process in a decentralized federal state.

ORGANIZATION OF MAB-6 ACTIVITIES

The early history of the Swiss MAB-6 programme is described by Freiburghaus and Zimmerman (1984). This report, together with interviews (see Author's Preface and Acknowledgements), forms the basis for this section. In 1973, the Swiss National Council (*Bundesrat*) determined that the Federal Office for Environmental Protection (FOEP) should establish a Swiss National MAB Committee. The committee first met in early 1974, and began substantive actions towards MAB-6 research in 1975. At this time, the committee had ten members: five natural scientists and five representatives of government agencies. The FOEP took responsibility for administrative matters and financed a part-time co-ordinator (who later became the full-time co-ordinator) who began publication of a *MAB-Information* newsletter, which continued until 1979.

In 1975, over 1,000 members of the Swiss Association for Environmental Research were informed of the proposed project. The 350 who expressed interest were sent questionnaires, to ascertain the extent of both basic scientific research and interdisciplinary studies on the relationships between human and natural systems in the Swiss Alps. Eighty questionnaires were returned, providing the basis for identifying potential test areas for research to fulfil the aims of the MAB-6 programme. Initially, twelve potential areas were identified. In 1976, the National Committee selected four of these as test areas: Aletsch, Davos, Grindelwald and Pays d'Enhaut (Figure 4.1). The following criteria were used:

(1) Level of ongoing or completed research relevant to the themes of MAB-6: a high level would provide a stronger basis for interdisciplinary research;

(2) Spatial distribution: representation from the east and west and the French- and German-speaking parts of Switzerland and from the outer and inner Aps;

(3) Accessibility to research institutions: to permit participation by scientists from as wide a range of institutions as possible; and

(4) Increasing impacts and conflicts resulting from current land-use and management practices.

As the number of interested scientists was rather large, working groups were set up to define research priorities for each test area. These groups consisted mainly of natural scientists from research institutions and some private companies, as well as practitioners, but few social scientists. A total of 105 ongoing, planned, or likely projects relevant to the overall MAB-6 theme was identified. However, these projects were very heterogeneous, and few of the members of working groups were used to interdisciplinary research. Consequently, the National Committee decided that projects had to be undertaken in an integrated manner and, in late 1976, supported the idea of a feasibility study in the Grindelwald area, which was financed by the Swiss National Foundation (SNF). The study began in spring 1977, with the aims of defining entire and partial systems, identifying interactions between the components and designing a digital terrain model and computerized information system. This approach was based to some extent on the experience of the Austrian Obergurgl MAB-6 project.

At this point, funding for the MAB-6 programme had not materialized.

However, the *Bundesrat* had approved funding for a national research programme on regional problems of mountain and marginal areas in 1976, and an expert group for this programme was established by the SNF. In summer 1977, the National MAB Committee asked the SNF to approve extra funds to integrate the forthcoming MAB-6 programme into this developing national research programme. This led to the constitution of a MAB sub-group of the expert group, including the most active members of the MAB committee, federal agency officials and other scientists. After consultation, the expert group proposed to the SNF that the MAB programme should be supported as an independent national research programme. This proposal was accepted in late 1977, after considerable debate on the scientific value of such an interdisciplinary project. However, the requested 5-year budget was only partially approved, and for only three of the four test areas, not including Davos. Provision of the remainder of the funds was to be dependent on a report to the SNF after the first 2 years of research.

As with all national research programmes in Switzerland, an expert group was established to provide a strong structure for decisions regarding the objectives, methodology and financing of research projects. The group mainly comprised the scientists from the National MAB Committee, which effectively ceased to operate from this time. The expert group was highly multidisciplinary, with natural and social scientists and practitioners as well as a representative of the federal administration, and met every 3 to 4 months throughout the project. Each member of the group had a specific task. In addition, a project leader and a full-time co-ordinator, both from the Geographical Institute at the University of Bern, were appointed. The project leader (who was also the chairman of the National MAB Committee) was not allowed to vote in the expert group's decisions.

The first activity of the expert group and co-ordinators was to prepare a plan which outlined the objectives of the programme and the expected areas of research (*Ausführungsplan*). This was widely disseminated in early 1978, and led to the submission of many project proposals. In general, those that were accepted came from the people who had been involved in the preparatory stages of the programme. Another early event was the expert group's identification of co-ordinators for the research in each test area. For each area, there was a more senior co-ordinator, who had previously worked there, and a younger scientist responsible for general project management.

MAB-6 RESEARCH ACTIVITIES

The Swiss MAB programme on 'socio-economic development and ecological carrying capacity in mountain regions' began in 1978. Its conceptual framework was based on three principal considerations (P. Messerli, 1984):

(1) That there are a number of problems common to all Swiss mountain regions, which need to be solved within the specific natural, socio-economic and cultural contexts of each community;

(2) That the level of decision-making relevant to ecological conditions is local or individual, so that these levels are of greatest importance for integrated mountain management; and

(3) That, because the preservation of the natural environment is no longer ensured by adapted agricultural systems, type and intensities of land use have to be defined in relation to ecological constraints and changing socio-economic needs.

This framework was concretized in a model of a regional socio-economic system, showing local- and regional-level linkages (Figure 4.2). The intention was to describe and analyse these linkages over recent decades, in order to provide information to:

(1) Achieve a better understanding of relationships between socio-economic development, land-use practices and conditions and availability of natural resources; and

(2) Attain a better integration of individual problems, and approaches to their solution, within the context of the systems under review.

At the beginning of the project, a very detailed list of variables was prepared, with the intention of assembling standardized data sets for the four test areas. However, this approach was not accepted by the test area co-ordinators, and the expert group decided that there should be no attempt to standardize either research approaches, or the level or means of integration of local populations and decision-makers, across the test areas. There were a number of reasons for this lack of standardization, stemming not only from differences in the experience and constitution of the multidisciplinary research groups and in the perspectives and political situation in the communities of the test areas, but also from the expert group's interest in assessing the value of different models for treating the topics considered in the programme.

In each project, one or two models (computer-based except in the Aletsch area) were used to support interdisciplinary research through the need to operate in a consistent framework and have complementary quantitative data inputs. While there was relatively little co-ordination between the projects in the early stages of field research, the degree of co-ordination increased at later stages, in order to obtain comparable results at the thematic level and to compare the applicability of the results on similar

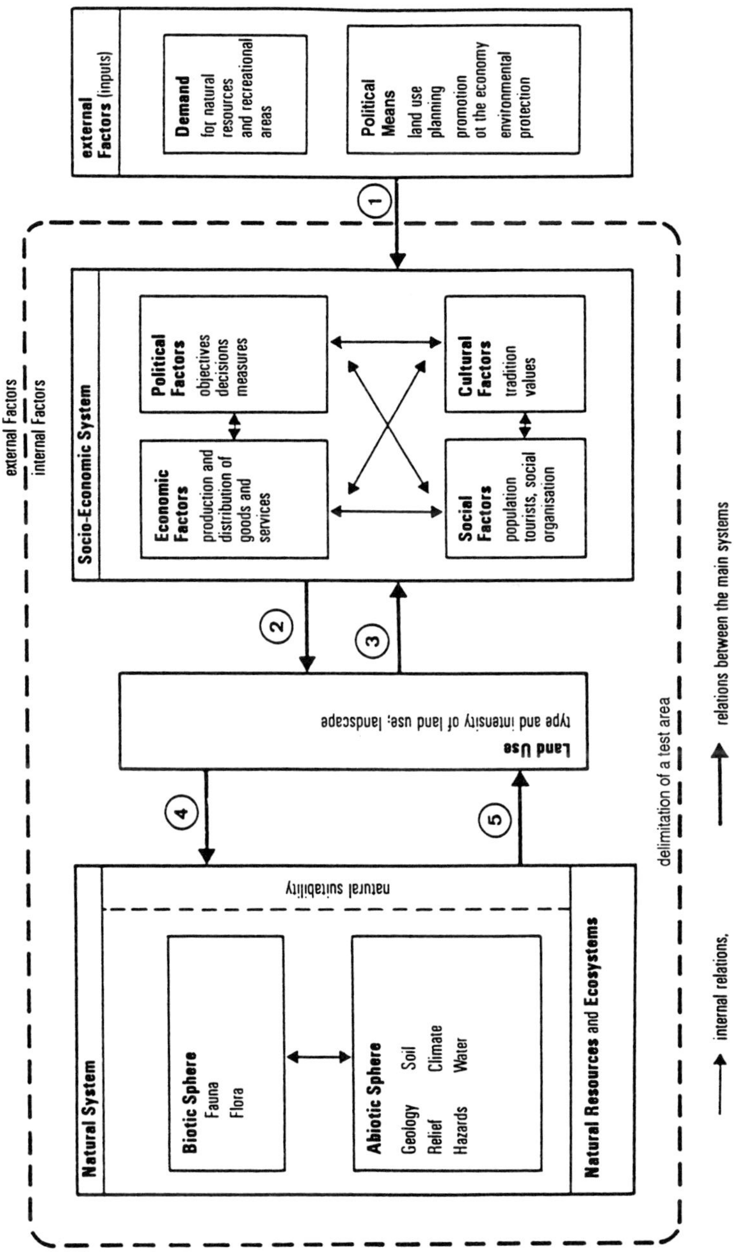

Figure 4.2 Schematic presentation of a regional socio-economic ecological system (from Messerli and Messerli, 1978, with permission)

topics from different models. Both during and at the end of the programme, some synthesis took place, with the aim of providing information for decision-making at all levels from individual to national.

The results of the programme have been disseminated through a wide range of scientific and popular publications. Only the major reports and publications are cited below; a full bibliography may be found in the synthesis book (Messerli, 1989). Links to the media were explicitly fostered, for instance through a short training course in 1983 (Küchli et al., 1983), and many Swiss scientific meetings – as well as the international meetings mentioned in the introduction – concentrated wholly or partially on the MAB-6 programme. There were two primary report series: the twenty-seven technical reports (*Fachbeiträge zur schweizerischen MAB-Information*) and the thirty-eight final reports (*Schlussberichte zum schweizerischen MAB-Programm*). Another important publication, early in the programme, was an edited book on *The transformation of Swiss mountain regions*, published in German and French (Brugger et al., 1984a) and partially translated into English (Brugger et al., 1984b). This was a joint production of the Swiss MAB programme and the national research programme on 'Regional problems in Switzerland'.

The remainder of this section considers the work done in the four test areas and concludes with a description of the thematic studies and syntheses.

Aletsch

The Aletsch test area (6,738 ha) is on the north side of the deeply-cut glacial valley of the Rhône in the Canton of Valais. It includes eleven communes, many of which reach from the valley floor to the alpine zone. The main valley settlements are located where the valley floor (720–1,100 m) is locally widened. The forested slopes are mainly steep and dissected by deep stream valleys. Most of the old permanent settlements are on 'sun-slopes' at altitudes from 1,170 to 1,360 m. Above 1,900 m is the flatter *Alp* (pasture) zone and the tourist settlements of Bettmeralp and Riederalp. Above this rises a steep and rocky ridge, culminating in the summits at altitudes up to 2,926 m. North-west of this ridge is the steep slope covered by the near-natural Aletsch forest and, below this, the Aletsch Glacier (Figure 4.3).

Until the 1950s, the area's economy was based on primary activities. The land-use pattern was typical of such areas, including three zones: the *Heimgüter*, with permanent settlements, fields and winter pastures; the subalpine *Maiensäss*, with irrigated pastures and fields and periodically-inhabited settlements; and the *Alp*, with summer pastures and buildings used only during the summer grazing season. These complementary zones were used in the typical pattern of mountain farming, with hay-making

Figure 4.3 Aletsch test area, showing the village of Mörel in the valley bottom, the ski resort of Riederalp, and the Aletsch glacier in the distance (photograph by U. Wittwer)

and farming in the lower two zones and domestic animals moving from the permanent settlements to the *Alp* for the summer (Messerli et al., 1980; Figure 4.4). While the economy primarily depended on primary activities, the limited resource base of the area was reflected in relatively small changes in population.

Changes in the socio-economic structure of the area were described by Mattig and Zeiter (1984) as part of the MAB Project. Summer tourism began in the area in the 1850s, but remained very limited until the late 1950s, in spite of improvements in road and rail access. From 1950, the basis of the economy changed rapidly from primary activities to tourism.

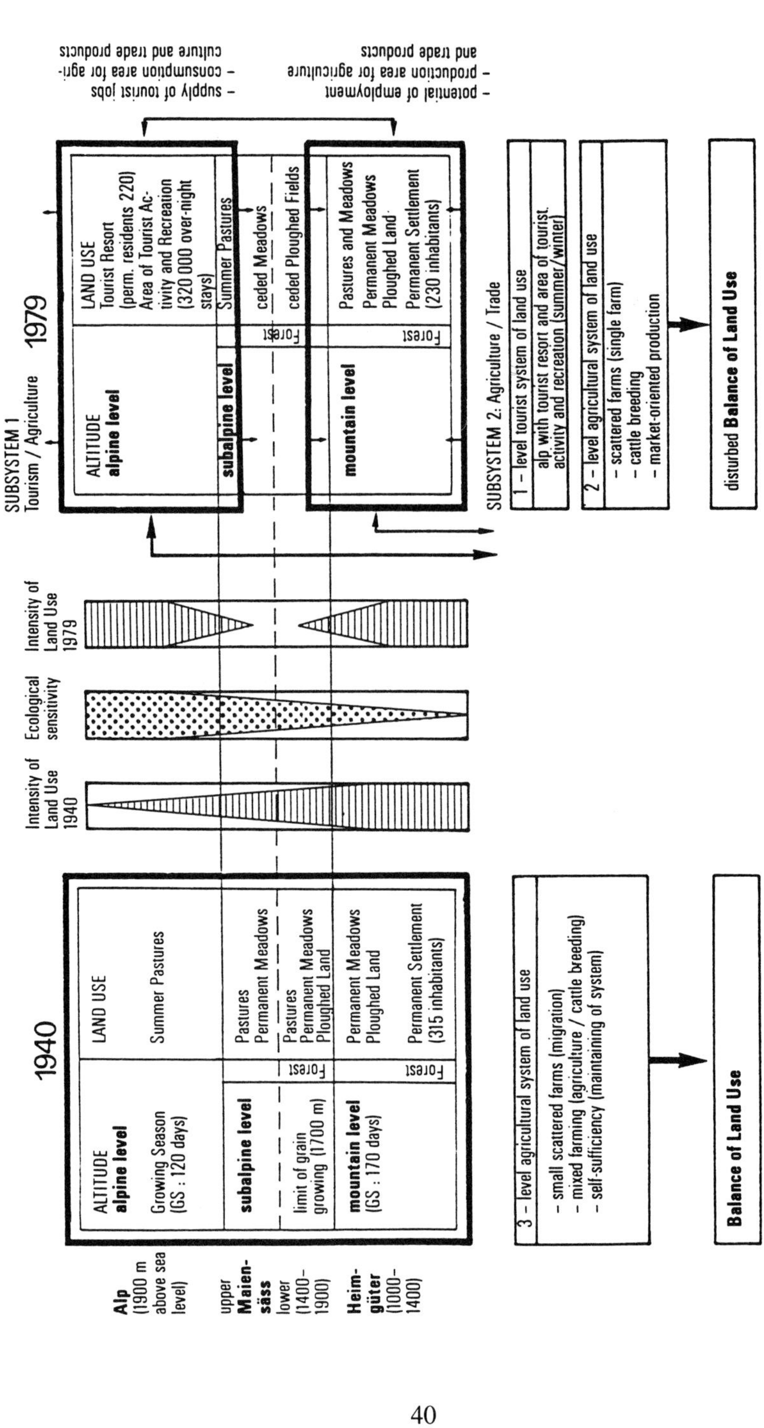

Figure 4.4 The consequences on land use, economy and society resulting from the change from the closed to the open system in the Aletsch area (from Messerli et al., 1980, with permission)

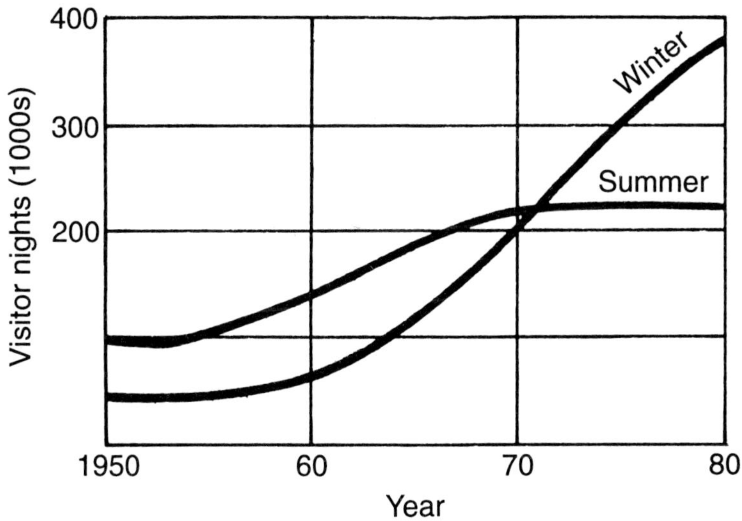

Figure 4.5 Trends in the numbers of summer and winter overnight stays in the Aletsch area. 1950 to 1980 (from Messerli, 1989, based on Mattig and Zeiter, 1984, with permission)

The impetus to this change was the construction of cable-cars to the *Alp*, followed by the construction of the new resorts of Bettmeralp and Riederalp. Demand for tourism and the facilities related to it rose rapidly into the mid-1970s, when winter became the primary season (Figure 4.5). By the mid-1980s, accommodation (largely self-catering) had been constructed for nearly 15,000 people, and there were about 1.2 million overnight stays a year. The population had grown again after declines in the 1940s and 1950s, and the labour market had become dominated by the construction and service sectors; the primary sector had shrunk to about 10 per cent, although many people worked in agriculture and tourism.

The integrated three-level land-use system now barely exists (Figure 4.6). The *Alp* – with ski-lifts, hotels and chalets – is the dominant zone; the meadows and pasture of the *Maiensäss* are abandoned; the arable land of the *Heimgüter* has been invaded by forests; the emphasis of the remaining farmers has changed from mixed farming to market-orientated cattle-breeding; and the forests are hardly used. The valley settlements have grown substantially in area, but without great population growth; most of the new construction has provided second homes and condominiums for short-term visitors.

The MAB research in the Aletsch area was co-ordinated by the overall project's full-time co-ordinator at the Institute of Geography at the University of Bern. The first projects started in early 1979. While some

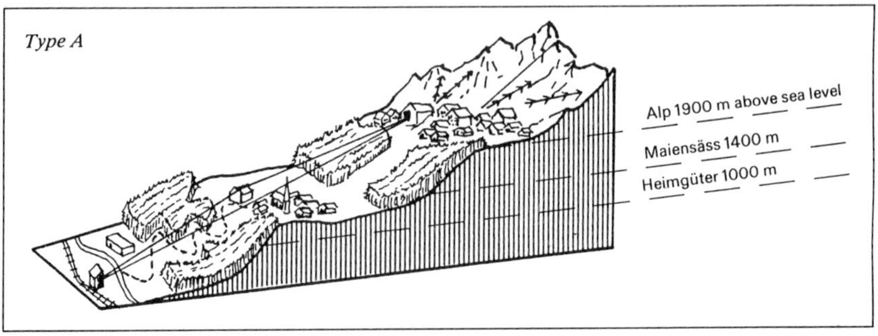

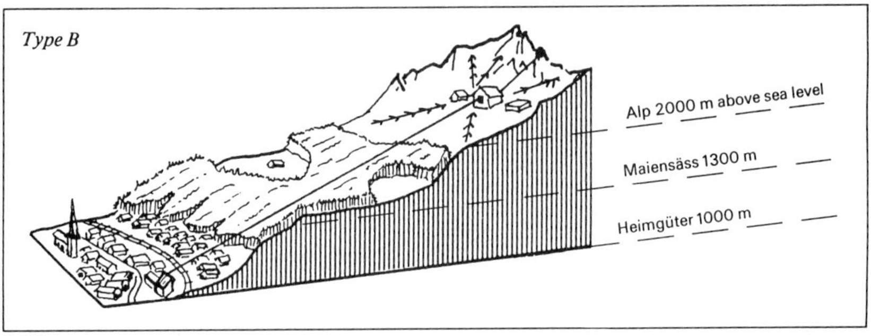

Figure 4.6 Two examples of the vertical organization of land-use zones in the Aletsch area. In type A, the original economic unit, with the permanent village on the sun-slope, has been divided into two sub-systems, with a ski resort in the upper *Alp* zone (e.g. Bettmeralp). In type B, the permanent village is in the valley and a cable-car has opened access to the *Alp* zone, where a ski resort has developed (e.g. Mörel and Riederalp) (from Messerli et al., 1980, with permission)

research was undertaken by scientists from the Federal Polytechnic (ETH) in Zürich, most was done by scientists and students (often as a *Diplomarbeit*: master's thesis) from the Institute, beginning with a field camp in 1979. However, local people were also involved, and provided good contacts to the local authorities. There is also a good field base, the Aletsch Centre of the Swiss Federation for Nature Protection, at Riederalp, whose director was involved in the MAB Project and further assisted in developing awareness of its results in the resorts of Bettmeralp and Riederalp.

With the exception of Mattig and Zeiter's (1984) research on the socio-economic system, and studies of agriculture (Nachbur, 1983; Staub, 1980) and landscape quality (Adamina, 1982), most of the research considered natural scientific themes, with varying degrees of emphasis on the influences of human activities. These included studies of vegetation and landscape, with particular emphasis on environmental protection (Beguin and Theurillat, 1982, 1987; Meessen and Luder, 1987), ornithology (Renevey, 1984), soils (Krause, 1982; Liniger, 1983), and forests (Bellwald and Graf, 1985).

Figure 4.7 Davos test area, showing the resort of Davos, which has occupied much of the best agricultural land, and the protective forests above (photograph by T. Günter)

While the MAB Project has generally had little impact on the people and authorities of the older, lower (slope and valley) settlements, the lines of communication between MAB scientists and the authorities of Bettmeralp and Riederalp led to the establishment of a permanent planning group after the project ended. This group used many of the results of the project in the development of the master plan for the two resorts in the early 1990s.

Davos

The Davos test area (17,130 ha) comprises the entire commune of Davos in the Landwasser, Flüela, Dischma and Sertig valleys (*Täler*) in the Canton of Grisons. The principal tourist settlement of Davos and smaller settlements are located on the flat valley bottom (average elevation of 1,600 m) of the main Landwassertal. The three side valleys contain small farming settlements. The sides of both the main and side valleys are steep, becoming flatter at *c.* 2,000 m at the edge of the *Alp* zone. Above this is the rocky alpine zone, culminating in summits from 2,500 to 3,146 m (Figure 4.7).

From the late twelfth to the mid-nineteenth century, Davos was a predominantly agricultural area, the fortunes of which fluctuated with mining booms (late sixteenth and early nineteenth centuries) and the development of trading routes. By the mid-nineteenth century, the agricultural sector was declining and much of the trade on which Davos depended had been diverted elsewhere. However, a new external source of income appeared in the 1860s, with the completion of a road to Landquart, in the Rhine valley: tourism, whose history is described in the MAB report by Kneubühl (1987). Most of the first visitors came for medical reasons; Davos was an early health resort for people with rheumatism and tuberculosis. Yet people did not come only for their health; facilities for both winter and summer sports were developed from the late 1860s.

The completion of the railway from Landquart to Davos in 1890 led to further development of the tourist industry. This was linked to major changes in demography – the population grew fivefold from 1870 to 1910, when 48 per cent were of foreign origin – and agriculture. The number of farming enterprises decreased, not only because of new opportunities for employment and the need for building land, which increased land prices, but also because of the sanatoriums' requirements for large volumes of milk. This resulted in a change of agricultural emphasis from cattle-breeding and arable farming to more intensive dairying. By 1900, employment in agriculture was only slightly more important than that in the service sector.

The tourist industry continued to grow rapidly until the First World War. During the interwar years, while tourism remained dominant, the number of visitors fluctuated greatly. Winter sports became increasingly important, especially after the construction of the first mechanical lifts in the early 1930s. During the Second World War, there was a strong demand for accommodation for refugees, the wounded and even tourists. In the late 1940s and 1950s, demand again declined, due to restrictions on the travel of potential foreign guests and the introduction of antibiotic therapy as a cure for tuberculosis. During this period, there was an overall decrease in the population and the sanatoriums were converted to 'sport-hotels'. From the late 1950s to the mid-1970s, numbers of visitors increased again (until 1973), and accommodation and downhill skiing facilities were expanded. Since the 1950s, winter has been the primary season.

Changes in population and employment since 1960 reflect the changing demand for tourism in the largest tourist resort in the Swiss Alps – with around 2.5 million overnight stays a year – as well as an important international commercial and conference centre. While the population grew by 7 per cent from 1960 to 1970, growth almost stopped after the demand for tourism declined after its peak in 1973. The service sector has continued to expand its dominance of the economy, and the agricultural sector now accounts for less than 5 per cent of employment. There is continued pressure for the development of the remaining agricultural land, still principally used by dairying operations, which have increased significantly in number since the mid-1950s. Most farmers also work in the tourist industry, which is often their primary source of income.

MAB-6 work in the Davos area was co-ordinated from the Federal Institute for Forestry Research (EAFV), in Birmensdorf near Zürich. After funds were approved in late 1980, it was decided that the research should focus primarily on relationships between the natural environment and land use, and that the framework for research should be a grid-based (raster) GIS, with a minimum grid size of 50 m × 50 m. The GIS/image processing system available at the ETH was relatively simple (by today's standards), with little potential for modelling, but provided facilities for visualization and superposition of different data layers to create thematic maps.

The proposals that had been submitted earlier were reconsidered, and decisions about their inclusion in the MAB project were made in relation to their potential for developing the GIS database and their proponents' willingness to contribute to the general goals of the project. At the same time, participants were required to work to the usual standards of their disciplines, even if this meant that their results had a greater degree of detail than required for inclusion in the GIS. Preliminary work began in 1981, and the main research projects in early 1982. While most of these were undertaken by scientists from the EAFV, the ETH and the University of Zürich, scientists from institutions in other cities were also involved, such as the Universities of Basel and Bern, the National History Museum in Chur, and the Snow and Avalanche Research Station at the Weissfluhjoch, above Davos. Communication between the scientists was facilitated by meetings and also by the fact that scientists doing field-work stayed in the same apartment in Davos. In addition, the use of the GIS also improved interactions within the project team, leading to substantive interdisciplinary discourse.

A number of publications deriving from the project consider the biophysical environment and human influences in the test area, with respect to soils (Krause, 1986), forests and their management (Hefti and Bühler, 1986), wildlife and their habitats (Müller et al., 1988) and air quality (Moser, 1985). Other reports describe changes in land use, and nature and landscape protection, in relation to tourism (Günter, 1985; 1987), and the GIS-based simulation model and its application for soils, vegetation, avalanche hazards and game (Binz and Wildi, 1988). An edited report with many maps summarizes all of these studies together with other research undertaken in the project (geomorphology, natural hazards, climate, vegetation) and provides brief recommendations (Wildi and Ewald, 1986). Figure 4.8 shows some of the results of this largely biophysical research, on altitudinal variations in settlement, forest cover and soil quality.

People living in Davos were informed about the project from its beginning, particularly through the local newspaper. However, Davos has

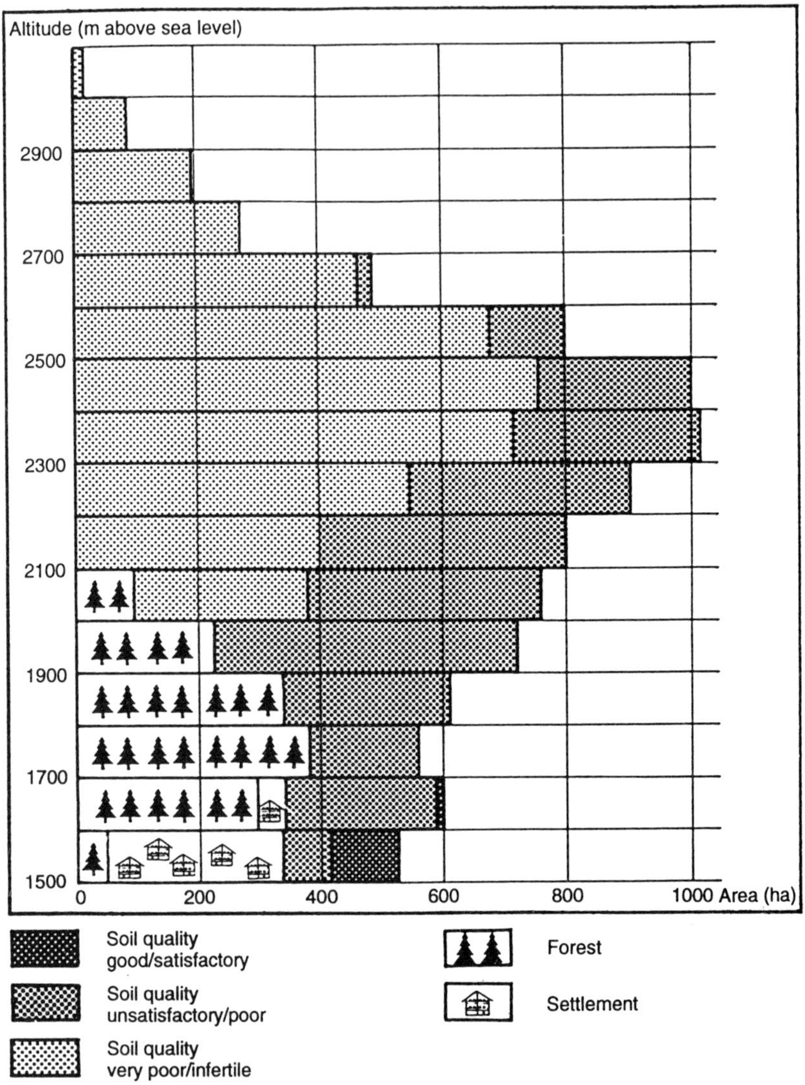

Figure 4.8 Distribution of land-use and soil quality zones by altitude in the Davos area (from Messerli, 1989, based on Wildi and Ewald, 1986, with permission)

many research institutions and newsworthy events, so there was initially little interest. Nevertheless, the Davos Association for Natural History Research was supportive, and provided important connections to the local administration. Local officials attended many of the presentations of intermediate and final results of the project. The primary stimulus to public interest came from the results of the survey of air quality, which revealed considerable pollution. As a result, a permanent air quality monitoring network has been established. One of the main results of the MAB project was a recognition of the importance of agriculture in the landscape, which has been recognized in the recent master plan for Davos, which used many of the results of the MAB research. A booklet summarizing the results of the project was recently published by the Commune of Davos (Frank, 1991). The GIS, which has now been transferred from a mainframe computer to a Macintosh microcomputer, has been used in many research projects since the MAB project ended.

Grindelwald

The Grindelwald test area (25,417 ha) comprises the catchment of the Lütschinen in the Berner Oberland, in the Canton of Bern. It contains one main settlement: Grindelwald, with houses scattered around the central core at altitudes from 950 to 1,200 m. South of the town are the steep rock walls of – from north to south – the Wetterhorn (3,700 m), Eiger (3,970 m), Mönch (4,099 m) and Jungfrau (4,158 m), and the two Grindelwald glaciers, almost reaching the valley bottom. The rest of the valley generally has forests on higher-angle slopes up to timberline at *c.* 1,800 m and, on lower-angle slopes, fields and pastures up to the passes of the Grosse Scheidegg (1,962 m) to the east and Kleine Scheidegg (2,061 m) to the west (Figure 4.9).

During the MAB project, both the long-term evolution of agriculture and settlement (Aegerter, 1983) and the development of tourism (Wiesmann, 1986) were investigated (Figure 4.10). The first hotels were built in the early nineteenth century. There was a rapid increase in demand for both summer and winter tourism from 1886 until the First World War, with construction of much accommodation and railways for both access and sightseeing. However, the population did not increase greatly. As early as 1900, about 40 per cent of farmers' earnings came from tourism, and strong links between agriculture and tourism have continued to develop; many farmers, while remaining active in the agricultural and forestry co-operatives that have existed since the fifteenth century, have hotels and small industries.

Tourism continued during the interwar period, which was marked by annual increases in winter visitation, while summer tourism fluctuated greatly. As in Aletsch and Davos, tourism increased rapidly from the mid-1950s to the mid-1970s. Much accommodation, especially self-catering, was constructed, as well as many ski-lifts (Figure 4.11). Overnight stays stabilized around 1 million a year from the mid-1970s, with a small dominance of summer over winter. However, in contrast to the other areas, Grindelwald has many day-visitors throughout the year because of its easy access to Bern and other major cities. While tourism is clearly the dominant

Figure 4.9 Grindelwald test area, with the Wetterhorn (3,700 m) in the background and the Grosse Scheidegg pass to the left (photograph by P. Messerli)

economic sector, on which 90 per cent of total incomes depend, it is difficult to specify the relative importance of other sectors, as many permanent inhabitants have multiple jobs. From 1955 to 1980 agriculture changed considerably, with the number of agricultural operations decreasing by 39 per cent and of full-time jobs by 70 per cent; but the number of part-time farmers increased by 40 per cent. In spite of these major societal changes, the very complex social structure – involving about 120 groups concerned with agriculture, forestry and tourism – has persisted.

The MAB project in Grindelwald was co-ordinated from the Institute of

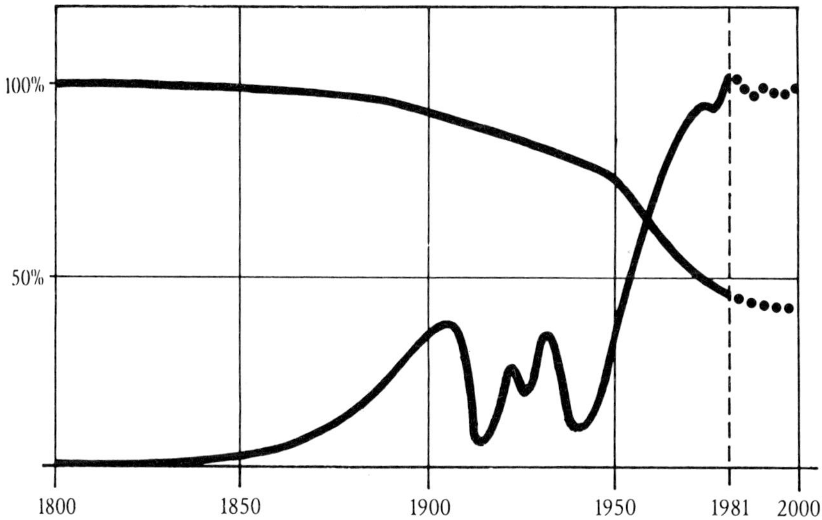

Figure 4.10 Comparison of the relative importance of agriculture (upper curve from left: number of enterprises) and tourism (lower curve from left: number of overnight stays) in the Grindelwald area from 1800 to 1981 (from Messerli, 1989, with permission)

Geography at the University of Bern, whose scientists and students undertook most of the work. In addition, scientists from the ETH and University of Zürich participated. Some field-work began in summer 1978, with most of the projects starting the next summer. Three models were used in the research. The first of these was a rudimentary GIS, based on polygons with homogeneous characteristics, whose development had begun during the 1977 feasibility study. This provided an overall framework for the research (Steiner and Zamani, 1984). The other models were a system-dynamic model and an input–output model, which were compared during the project (Apel, 1983).

The research led to the preparation of a large number of theses and publications. There are many with an emphasis in the natural sciences: on hydrology (Leibundgut, 1987), soils (Scheurer, 1985), climatology and air quality (Coendet, 1979; Kunz, 1983), wildlife (Schiess, 1988) and forests (Kupfer and Langenegger, 1985). Many of these studies give significant consideration to the importance of these resources for human populations. Agriculture was a major focus of research, with studies on grassland vegetation and fodder production (Pfister, 1984), and history, economic and labour force structure, land use and effects of tourism (Naegeli-Oertle, 1986); and a detailed scenario-based evaluation of the availability and use of resources (Scheurer-Lietz, 1989).

As well as the major study on the economic, societal and spatial aspects of tourism (Wiesmann, 1986), there were other studies of this sector, considering ecological impacts on vegetation and wildlife (Langenegger and von Grünigen, 1983; Stucki, 1984), and access and traffic (Hoppler et al., 1985). In addition, an environmental impact assessment of proposals for holding Winter Olympic Games in Grindelwald was undertaken (Messerli et al., 1986). This special study, which used scenario techniques, was supported by the Canton of Bern. Other reports considered the different value systems of established residents, recent immigrants and tourists (Meyrat-Schlee, 1983) and the evaluation of aesthetics, for which a specific methodology was developed during the project and applied to both past and current landscapes of Grindelwald and other parts of Switzerland (Grosjean et al., 1986; Figure 4.12).

From its beginning, the local people and authorities were aware that the project was underway, but the research team decided to limit discussion of the project during the field research phase because of its complex nature. However, the publication of two articles in regional newspapers, presenting intermediate research results that showed that tourists would like better public transport facilities, and that local people were wary of the influences of immigrants, led to considerable interest in, and concern about, the findings of the project. Consequently, as the field research neared its end, results were presented and discussed at many meetings of local people. There was often good agreement between the scientific findings and local beliefs and attitudes. Two important instances were the recognition of the importance of agriculture for tourism and in maintaining the socio-cultural identity of the local population, and concern about the risks of the continued construction of tourist facilities, particularly second homes.

In the later stages of the project, a planning group was constituted, involving local administrators, representatives of interest groups, and the project co-ordinator. Over a period of a year, he took part in at least one meeting a week with local interest groups. The main purpose of these meetings was to try to resolve incompatibilities between goals, in order to develop draft guidelines for planning. Simulation models based on the MAB research were used to illustrate some of the arguments. A final summary report of the results of the project and its implications for policy and practice was prepared as a result of this process (Wiesmann, 1988).

The main conclusions of the MAB research and the many meetings was that the future development of tourism needed to be in a qualitative direction, and that a careful balance had to be struck between the interests of long-established and newly-arrived inhabitants and property owners. Over a period of a year and a half, beginning during the synthesis phase

Figure 4.11 (Opposite) Development of tourist accommodation (A) and transportation capacity (B) from 1940 to 1982 in the Grindelwald area (from Wiesmann, 1986, with permission)

(A) Tourist accommodation capacity

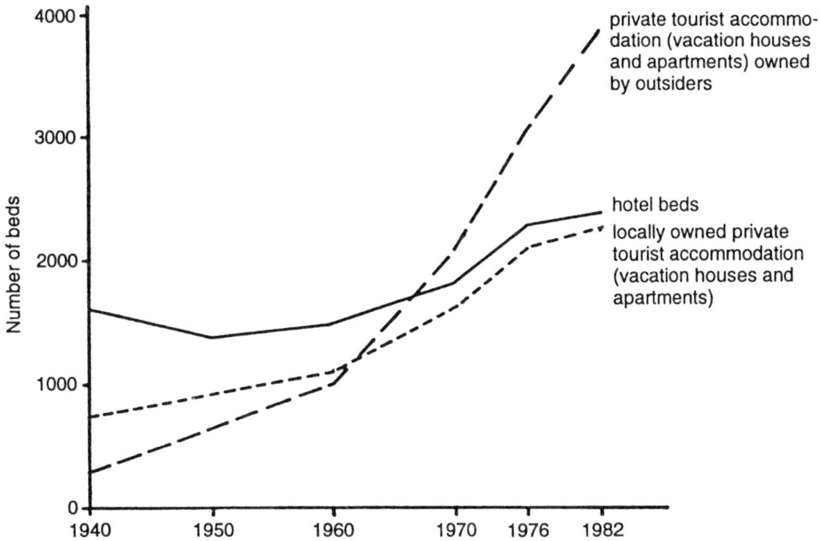

(B) Capacity of tourist transportation facilities

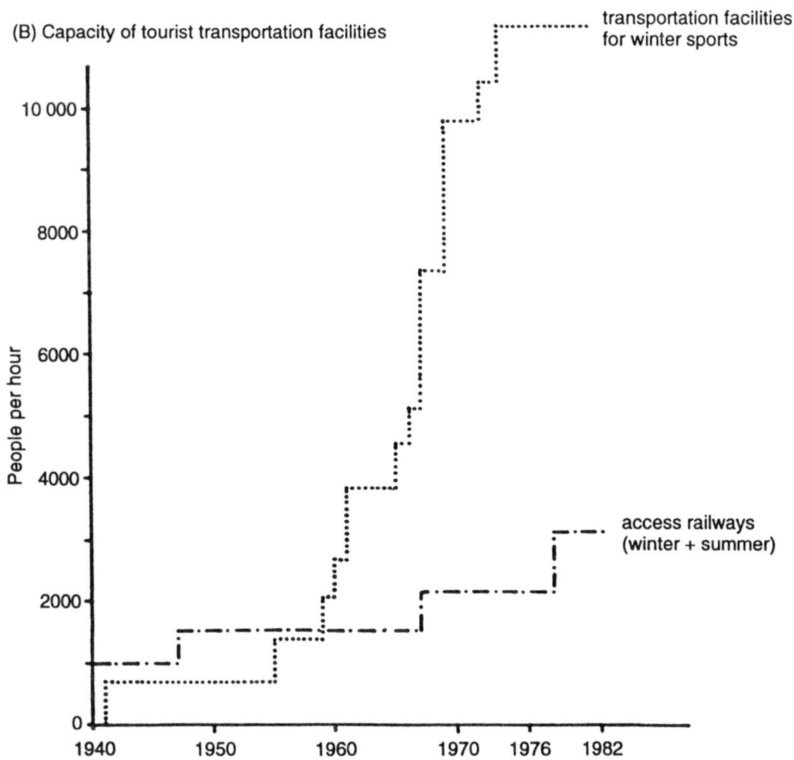

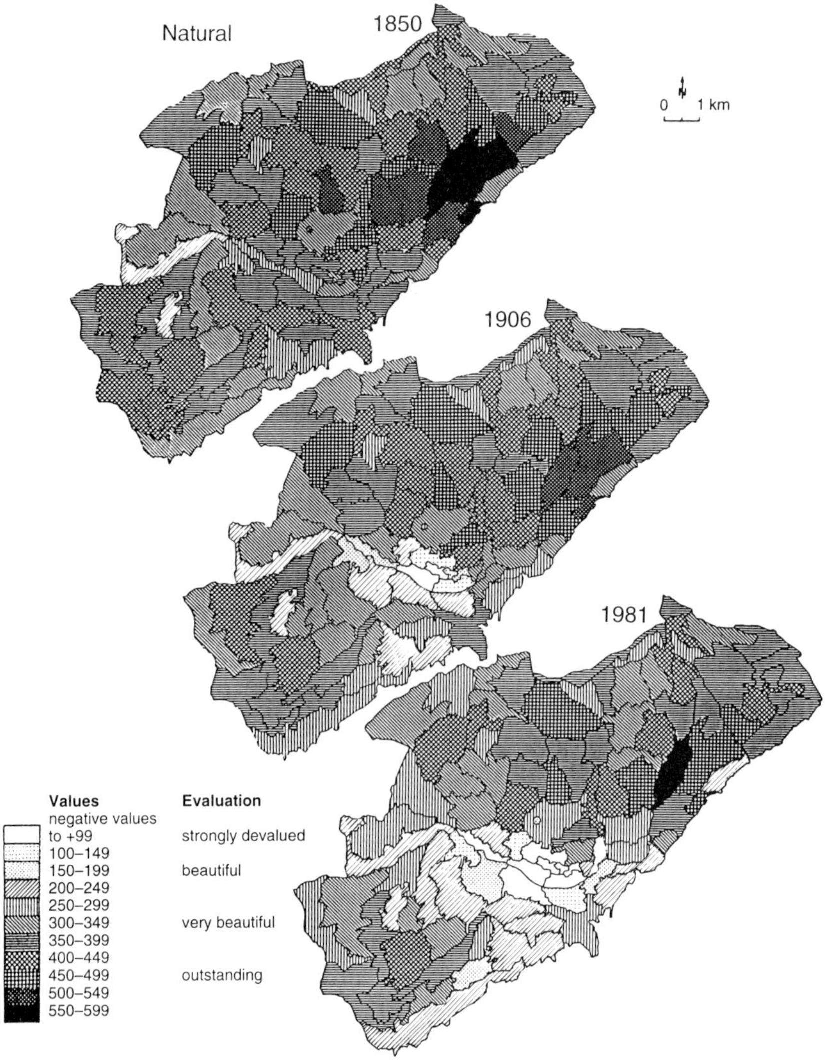

Figure 4.12 Aesthetic evaluation of the landscape of the Grindelwald area in 1850, 1906 and 1981: 'natural' type on the left, 'traditional' type on the right (from Messerli, 1989, based on Grosjean et al., 1986, with permission)

of the MAB programme, the planning guidelines were developed into a master plan, which was accepted by 90 per cent of the population in a vote in 1988. Some of the measures it included were a restriction on the construction of second homes, a requirement that 40 per cent of their area should be available for local people, a new transportation plan and

Figure 4.12 *Continued*

guidelines for agriculture and forestry. Advertising for the area began to stress the high quality of Grindelwald's physical and cultural environment. The planning group remains active.

At present, the local development plan is being reviewed on the required 15-year cycle, with considerable use of the GIS database, which has been transferred from a mainframe computer to the ArcInfo system on a personal computer. Since 1991, a number of studies have been in progress, mostly by students from the Institute of Geography of the University of Bern. Many of these studies are evaluations of the impact of the master plan: on construction and the renovation of old buildings, the success of the policies regarding second homes and strategies for maintaining the agricultural labour market and activities. Other studies use the GIS to monitor biophysical changes since the MAB field-work concluded in 1982. In addition, linkages between the economic sectors in various actor groups are being analysed. As noted above, most people in Grindelwald have multiple jobs. However, the research in the MAB programme was generally sectoral, so that the linkages in actor groups, which seem likely to have considerable implications for environmental management, were barely investigated. In summary, the MAB project was only the beginning of a process of applied scientific research which continues, with many positive benefits for the people of Grindelwald.

Pays d'Enhaut

The Pays d'Enhaut test area (18,608 ha) is in the pre-Alps of the Canton of Vaud, and includes three communes (Château d'Oex, Rossinière and Rougemont). The wide valley floor, with an average altitude of 1,000 m, is surrounded by forests, with alpine pastures above and mountains rising to 2,540 m (Figure 4.13). The traditional economy, comparable to that of the Aletsch area described above, began to change in 1903, when a railway was built into the area from Montreux on the Lake of Geneva. Tourism grew until the First World War, and again after it, but collapsed almost totally in the Depression of the 1930s, with widespread bankruptcies. From this time, the population began to decline, and also to age, as younger people emigrated; trends that intensified in the 1960s with changes in agricultural practices, mechanization and increases in the size of farms. However, the area has largely maintained its rural characteristics and employment in agriculture remains relatively high.

As described by Darbellay and Stucki (1984: the main source for the following discussion of activities before the MAB-6 project), in the late 1960s, developers began the construction of major tourist accommodation and skiing facilities in the area. Most of the accommodation was self-catering, built by and for people from outside the area. The local authorities were concerned about the implications of such development and established a commission, headed by the local-born director of the hotel school in Lausanne, to consider these issues. The commission included both local people and outsiders, and commissioned studies which considered agro-sylvo-pastoral topics, the potential for tourism development (particularly

Figure 4.13 Pays d'Enhaut test area, with Château d'Oex in the middleground (photograph by P. Messerli)

skiing), and a centre for secondary education. From 1969 to 1975, the studies were undertaken by scientists from regional institutions, often in close collaboration with local people. However, the development of tourism continued concurrently with the studies, leading to a tourist sector that occupied much good agricultural land and, because of its self-catering emphasis, provided relatively little employment for local people. Even now, only one-third of employment depends on tourism; far less than in the other test areas or the Swiss Alps as a whole.

In 1975, the federal law for investment assistance for mountain regions came into force. In order to apply for assistance, communes had to organize themselves on a regional basis and develop regional plans. Consequently, the three communes created the Association for the Development of the Pays d'Enhaut (ADPE), which asked various regional institutions, as well as the Institute of Rural Economy (IRE) of the ETH in Zürich, to undertake the work necessary for the regional plan. This involved scientific research and about sixty meetings in which over seventy local people took part, and led to an analysis of the current situation and potentials in 1977, followed by the definition of objectives and methods for development in 1978. Local people of all ages were involved throughout the project.

In 1979, the federal Department of Public Economy approved the regional plan. This was also the year when the MAB project began. It was co-ordinated through the IRE's new Centre for Rural, Mountain and Environmental Studies (CERME) in Château d'Oex. Participating scientists came from the Universities of Geneva and Lausanne, the federal agronomic research station at Changins, and the cantonal forest service as well as the IRE. Most of these institutions were represented in the scientific committee which co-ordinated the research: the *Direction du Projet MAB-Pays d'Enhaut* (DPMPE).

The process of defining the component studies of the MAB project had been complex. One reason for the choice of the Pays d'Enhaut as a test area was that there were many scientists who had worked on the previous projects, which had led to the creation of a good set of baseline information. These scientists and others interested in working in the MAB-6 project established a working group which also included local people, to define an interdisciplinary research project which considered not only topics of scientific interest, but also issues of local concern, such as the risks associated with an economy dependent on tourism and the results of establishing a nature reserve. Research within the project was to lead to the establishment of a detailed database, with a strong emphasis on socio-economic topics, that could be used for simulation, particularly through linear programming models. Furthermore, a MAB-Pays d'Enhaut Association (MPEA) was to be established and be responsible for the management of the project. However, the working group's proposal, particularly its management aspects, was not entirely accepted by the expert group. They approved most of the constituent research activities, but on condition that these were submitted by individual institutions, and that financial oversight remained with the overall project co-ordinator in Bern.

Although the expert group refused to allow an autonomous local association to manage the project, the MPEA was founded when it began in autumn 1979. It was run by a committee of six local people and three research scientists. With the objective 'to improve the exchange of information between research workers and the population,' it published a newsletter and held many meetings to strengthen links between participating scientists and local people. It also aimed to maximize the complementarity of the MAB-6 and regional development projects, recognizing that there was potential for conflict, as discussed at a meeting of the Swiss National Commission for UNESCO in 1980, attended by many local people.

Both linear programming (Fracheboud, 1984) and dynamic balance models (Kyburz and Schmid, 1983) were used during the project. Other research was divided into six themes, for each of which partial syntheses were prepared: agriculture (DPMPE, 1985a; Mercier et al., 1987), demography and economy (DPMPE, 1985b), forestry (DPMPE, 1988a), land-use planning and nature protection (DPMPE, 1986), society and environment (Lieberherr-Gardiol, 1984) and tourism (DPMPE, 1985c). The results of the partial syntheses were drawn together into a two-volume final report (DPMPE, 1988b) with ten chapters. The first two of these are introductory. They are followed by summary chapters on the natural environment, its use and management, and human impacts; demography and economic activities; an anthropological analysis of relationships between the population and the environment; potential conflicts between land uses and the natural environment; and possible futures, with various scenarios. The three final chapters present the overall methodology, conclusions and an evaluation of the project.

Throughout the MAB project, local people were kept informed of its progress through the newsletter and regular meetings. From an early stage, they were also involved in the development of simulation models and played an active part in discussions based on the results of model runs. In addition, from the beginning of the project, its proponents had recognized that the final results had to be presented in a form that local people could understand and utilize. Consequently, near the end of the synthesis phase, the MPEA invited a wide cross-section of local people to meetings at which results and simulations were presented and discussed. These meetings led to the publication of a book on possible futures for the Pays d'Enhaut (Lieberherr-Gardiol and Stucki, 1987). The MPEA was disbanded in 1987.

The MAB project appears to have been successful in transferring scientific knowledge into the decision-making process. However, because the regional planning process, co-ordinated by the ADPE, continued through and after the MAB project, it is difficult to unequivocally credit the project as the source of policies or changes in values. Nevertheless, it seems to have been instrumental in developing local awareness of the risks of excessive dependence on narrow economic sectors, such as construction or tourism – although demand for second homes began to decrease in the late 1970s for external reasons – and conversely in showing the benefits of maintaining a traditional agricultural landscape and a diversified economy. In recent years, one of the results has been a renaissance and growth of the co-operative production of the local cheese, which is sold at high prices both in Switzerland and abroad. The success of these ventures has permitted the participating farmers to opt out of the national agricultural subsidy scheme.

One should stress that not all of the policy-relevant findings of the MAB project, for instance in relation to nature conservation, were implemented in local decision-making. This is currently leading to some conflicts in the Pays d'Enhaut. While information obtained during the MAB project is often requested for specific local decisions and the simulations are still used for teaching, planned studies to monitor changes since the field research stage and evaluations of scenarios have not been undertaken. Nevertheless, the approaches developed during the MAB project strongly influenced the Spanish project in the Catalan Pyrenees, and have been widely disseminated by CERME both in Switzerland and abroad, through co-operative work with institutions in Italy, Japan, the United States and other countries.

Thematic studies and syntheses

As well as the reports concentrating on individual test areas, referred to above, a number of thematic reports were prepared during the Swiss MAB-6 programme. Some of these were conceptual, others were based on research in all four test areas and a few were on themes peripheral to the main programme, such as the environmental impacts of hydroelectric power plants (Aegerter and Messerli, 1981). An important early conceptual report was on ecological stability (Gigon, 1981). This theme was further explored at an international workshop on the stability and instability of mountain ecosystems, at Riederalp in the Aletsch test area, in 1981 (Messerli and Ives, 1984). It was also the topic of the last technical report, on traditional solutions, current problems and perspectives for the future of alpine cultural landscapes (Bätzing, 1988).

Most of the other thematic reports provided links from the MAB programme to regional-level issues. These latter included many of the early technical reports, which were integrated into the book edited by Brugger et al. (1984a,b) and will not be mentioned here, and also Wanner's (1983) report on social change in peripheral agricultural areas. Spring et al. (1986) also considered changes in agricultural systems, with particular attention to symbioses between agriculture and natural environments in past, current and future perspectives. The future of forestry, another essential component of mountain economic and ecological systems, was considered by Combe and Frei (1986), drawing on the experience in the four test areas to develop an empirical basis for guidelines for forestry management and policy throughout the Swiss Alps. This comparative analysis was extended to the international scale in Price's (1990b) assessment of the origins, implementation, and results of forest policies in the Aletsch and Davos test areas and two comparable areas in Colorado, the United States.

All of these studies, while focusing on agriculture and forestry, emphasized the importance of integrating tourism into considerations of the future of mountain systems. Two other reports specifically considered this issue: by Schwarze et al. (1983), on the relationships between tourism and recreation, on one hand, and agriculture and forestry, on the other; and by Müller (1986), on the implications of tourism in mountain communities, with particular attention to the development of tourism policy. In addition, Krippendorf (1986) published a book for the general public, based on the MAB tourism research, entitled *Alpsegen, Alptraum* (Alpine blessing, alpine nightmare), with the subtitle 'for the development of tourism in harmony with man and nature'. A final policy-oriented report was prepared by Schwarzenbach (1987), who presented guidelines for policies for nature, landscape and environmental protection in mountain areas.

As well as these reports, the programme co-ordinator published two syntheses of the overall Swiss MAB-6 programme: a review of findings and conclusions, focusing on models and methods (Messerli, 1986) and a more wide-ranging book, entitled *Man and nature in the Alpine living-space* (Lebensraum): *risks, opportunities, and perspectives* (Messerli, 1989). The report on models and methods (Messerli, 1986) contains five chapters. The first provides an introduction to 'system research between pretence and reality'. The second, on 'the Swiss MAB experiment: an attempt towards comprehensive analysis of a mountain ecosystem', presents the theoretical and conceptual bases of the Swiss MAB programme, covering reductionist, analytic and holistic approaches from the natural and social sciences. The third discusses the definition of problems at the regional scale; how these may differ from problems defined at larger scales; and presents the Grindelwald and Pays d'Enhaut simulation models, the Davos landscape model and their results. The fourth discusses the methods and findings of scenarios, with respect to development in Grindelwald and Pays d'Enhaut and land use in Davos. The final chapter gives a summary and conclusions.

The synthesis book (Messerli, 1989) draws on the work in the four test areas, as well as the thematic reports, to provide an evaluation of the results of the programme and its policy implications. Its six chapters are entitled: 'Man in the Alpine living-space, a constant battle to maintain ecological stability'; 'the alpine cultural landscape, the threatened basis of modern development in mountain regions', 'tourism and growth since the Second World War: the trigger of the "erosion" of the cultural landscape'; 'the dynamics of the development of tourism, as reflected in the MAB test areas'; 'a new orientation of development strategies in mountainous tourist regions'; and 'a new balance between agriculture, tourism, and nature'. Figure 4.14 shows the general model of interactions between the components of these systems. The book concludes with a summary of the main findings of the Swiss MAB programme in point form (in French, German and Italian) and a complete bibliography of the programme.

CONCLUSIONS

The Swiss MAB-6 programme was remarkable for the thoroughness of its planning phase and the diversity of research questions, approaches and results: scientific, societal, and policy-oriented. The choice of four test areas permitted a large population of Switzerland's research institutions, as well as other organizations concerned with the management of various resources, to

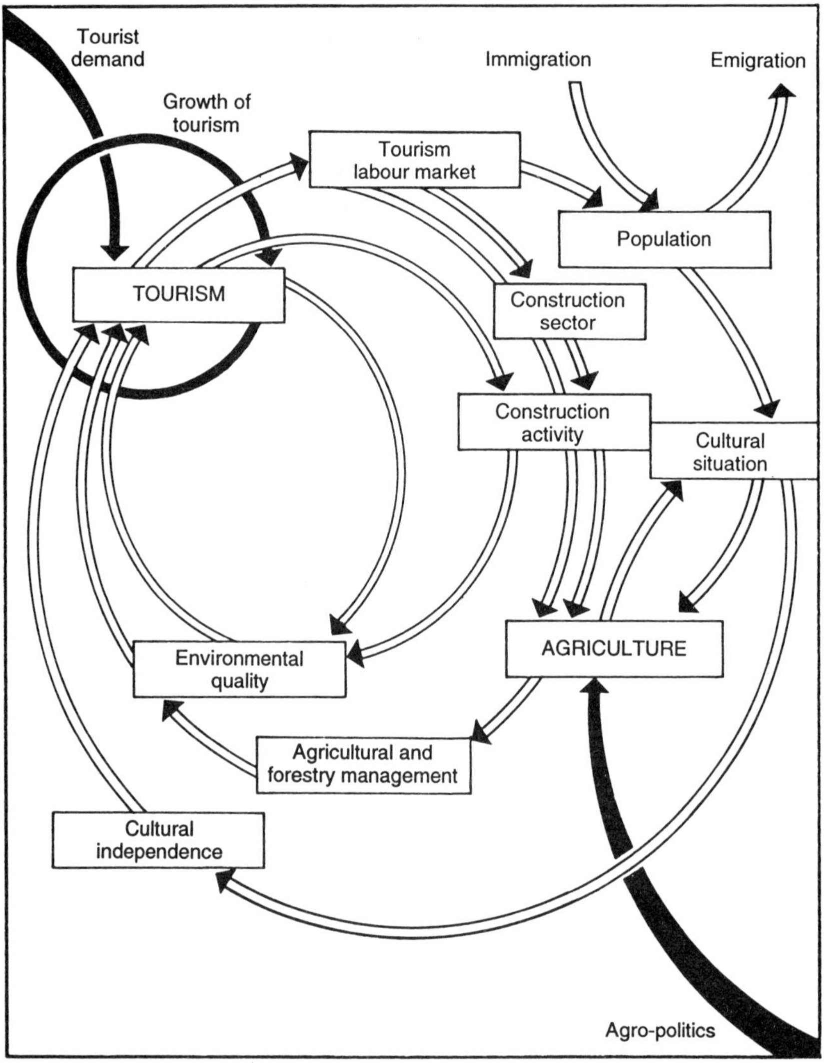

Figure 4.14 General model of growth-dependent relationships in the mountain agriculture–tourism–nature system (from Messerli, 1989, with permission)

participate in the programme. Nevertheless, the creation of functional interdisciplinary research teams was not easy. A number of factors which led to their success can be identified; strong leadership both from an expert group that was highly multi-disciplinary and committed, and from the overall and test-area co-ordinators; research themes and approaches that were feasible and interesting for participating scientists; and adequate funding, made possible by high-level political support, especially after the results of the first 2 years, which led to the release of the remainder of the funds and permitted the Davos project to go ahead.

In retrospect, it appears that the programme's diversity was one of its strengths. However, in its early stages, a considerable degree of standardization was anticipated. This would have permitted a greater degree of comparison, particularly at the quantitative level, between the results obtained from the four test areas. It might also have allowed the development of models at the regional 'control level' foreseen in the model of a holistic approach to the man–environment system, drawing on the experience of the Austrian Obergurgl project, which was developed early in the Swiss programme and elaborated throughout its course. Nevertheless, the expert group and overall co-ordinators recognized early that, for practical as well as intellectual reasons, the range of participating institutions and test areas would have to be reflected in a diversity of research themes and approaches. There was some flexibility in the choice of approaches, though this did not extend to the proposal of the Pays d'Enhaut scientists to have overall responsibility for the management of that project.

The four projects may be contrasted and compared according to a number of themes. The discussion below focuses on the emphases of the research and their local application. The concluding paragraphs consider the broader implications of the programme.

Both the Aletsch and Davos studies strongly emphasized natural science research and its applications for land use. However, the research approaches were rather different; in particular, while the Davos project was strongly influenced by the GIS approach, the Aletsch project did not have a specific research framework other than the programme's overall holistic model. In both cases, the results of the research have recently been used in developing master plans. Yet the master plan for the Aletsch area only concerns the upper settlements (parts of four communes), while the Davos master plan is for the entire commune.

It must be recognized that planning is inevitably more complex in the Aletsch area because it includes eleven communes. However, one weakness of the MAB project in this area was that it tended to concentrate on the slope communes and recreational zone and had little success in integrating the people and economy of the valley settlements. In contrast, the people of Davos became highly interested in the MAB project when high levels of air pollution were revealed. This was one stimulus to public realization of the need for better environmental management. Another was increasing awareness of the importance of improved forestry management, stimulated by the local forester after significant damage resulting from 'natural' causes but intensified by inadequate maintenance and harvesting; a problem also demonstrated by the avalanche hazard scenarios derived from the MAB research. In consequence, the development of a master plan was of interest to many local people, and the results of the MAB research, which had been compiled in a way that was highly suitable for planning, were of great value. Thus, over the long term, the GIS has proved to be valuable not only for research, especially evaluating changes in the biophysical environment, but also for planning.

Both the Grindelwald and Pays d'Enhaut projects utilized complex computer models for integrating research results and simulating possible futures, with emphasis on the integration of agriculture with other economic sectors, particularly tourism. However, the projects differed in many fundamental ways. The Grindelwald project, which was the most complex and wide-ranging of the four MAB-6 projects, developed a very detailed analysis of the man–environment system and its various components, integrated in a system-dynamic model. However, the field research was undertaken with little direct consultation with local people. Local awareness (and concern) about the project only developed when the first results became known through newspaper reports. Partly as a consequence, the synthesis stage was marked by a very intensive involvement of local groups in discussions with the project co-ordinator with respect to the future of the area. Simulation modelling was of great help to illustrate the implications of contrasting policy options. Eventually, an overall compromise was reached, based largely on the conclusions of the MAB project, and concretized in a master plan accepted by a large majority of local citizens.

In contrast, the Pays d'Enhaut project, building on local experience of applied scientific research over more than a decade, involved local people from the outset. The choice of research topics, with a marked emphasis on socio-economic and planning issues, partly reflected local interests. The formal designation of the unique Pays d'Enhaut MAB-6 Association was a strong catalyst to substantive information flow and discussion about research findings throughout the project, and led eventually to the publication of a book aimed at the general public. Simulation modelling, using linear programming, was developed at an early stage of the project, and used throughout it, to explore possible futures for the area. The project team was also well-integrated, as reflected in the joint authorship of the various partial syntheses and final report. Nevertheless, it is hard to evaluate the local implications of the MAB project, given that its findings and conclusions generally were in line with those of previous and concurrent regional research and planning.

In spite of the diverse research approaches and economic and cultural situations in the four study areas, there is considerable consistency in the conclusions of the MAB-6 projects. The importance of a traditional agriculture in maintaining not only the Alpine landscape but mountain communities was generally recognized, particularly in Grindelwald and the

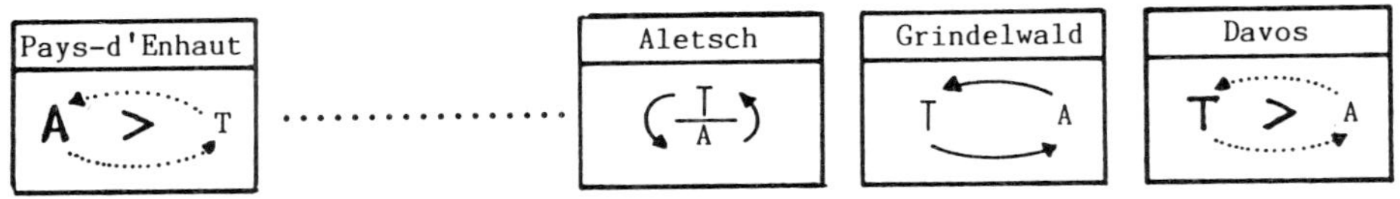

Dependence of labour market on tourism (Direct & indirect)	30 %	Increasing	80 %	90 %	With a service sector independent of tourism
Seasonality	Wi > Su		Wi >> Su	Wi < Su	Wi >> Su
Accommodation	PH >>> H		PH >> H	PH >> H	PH > H
Population	↘		↗	↗	↗
Problems of coexistence of agriculture and tourism	Integration of tourism development in a characteristically rural region	Prevention of the uncoupling of traditional occupations from the economic structure of tourism	Integration of traditional and modern farm types in the development of tourism	Further development of a long-term experience of co-existence of agriculture and tourism	Prevention of a further decline in agriculture to safeguard management of the available land area
Regional type	Agro-tourist periphery →				Tourist centre

Figure 4.15 A comparison of the four Swiss MAB test areas, showing the broad spectrum of types of regional tourism development from the agrotourist periphery to a fully developed centre which benefits from service functions independent of tourism: A = agriculture, T = tourism, Wi = winter, Su = summer, PH = parahotellerie (second homes and condominiums), H = hotels, economic dependence on tourism of the centres increases from left to right (from Messerli, 1987, with permission)

Pays d'Enhaut, in spite of very different histories and economic systems. This recognition has since been reflected in national policies which stress the importance of mixed agriculture, particularly in relation to medium- and long-term uncertainties of various origins. Other general conclusions were that overdependence on one or two economic sectors – particularly tourism and construction – is undesirable, and that qualitative approaches to tourism are preferable for both local communities and visitors. Such findings, stressing the importance of environmental and cultural quality and diversity, have been incorporated into national policies for tourism and a current long-term planning project, supported by the Federal Office of Tourism. The typology of community structures, developed on the basis of the comparison of the four test areas (Figure 4.15), is also being applied across the Alps, and may provide a basis for future work within the scope of the Alpine Convention signed in 1991 by all of the states of the Alps.

Outside the realm of policy-making, the Swiss MAB-6 projects have led to advances in the theory and practice of human–environment research in other mountain regions. The overall model, developed early in the programme, was incorporated into the framework of the German MAB-6 project in Berchtesgaden and Soviet MAB-6 research in the Caucasus. The Pays d'Enhaut approach, involving local people from the planning stage and throughout the project, was adopted by the Spanish MAB-6 project in the High Catalan Pyrenees. This approach, and the simulation modelling methodology developed during the project, have also been disseminated widely by CERME. Similarly, the approaches developed by the two project groups based at the Geographical Institute at the University of Bern have been applied in other countries. In at least one case, this has led to a realization that some of the MAB-6 research was inadequate. Thus, research methods to assess the environmental effects of actor groups involved in multiple economic activities, developed in Kenya, are now being applied in Grindelwald; where an evaluation of the impacts of the 1988 master plan, derived largely from simulation models developed during the MAB-6 project, are also underway. In summary, the Swiss MAB-6 programme, although largely fulfilling its objectives and completed within its specified timeframe, was not merely a discrete group of research projects, but also a catalyst to greater understanding and many other activities and actions relating to the concerns of MAB-6 in the Swiss Alps and other mountain regions.

CHAPTER 5

GERMANY

MOUNTAIN REGIONS

Most of the mountains of Germany (Figure 5.1) are middle mountains (*Mittelgebirge*). These include areas of high relief both in the centre of the country – such as the Harz (highest peak: Brocken, 1,142 m), Thüringer Wald (Beerberg, 982 m) and Rhön (Wasserkuppe, 950 m), which includes a biosphere reserve designated in 1991 – and those near and on its borders with Ausrtria, the Czech Republic and France. The border mountains include the Black Forest (*Schwarzwald*: Feldberg, 1,493 m) in the south-west, near the border with France, and the Hercynian mountains along the eastern border with the Czech Republic. These are, from north to south, the Ore Mountains (*Erzgebirge*: Fichtelberg, 1,204 m), the Fichtelgebirge (Schneeberg, 1,051 m), and the Bohemian Forest (*Böhmer Wald*: Grosser Arber, 1,457 m). The latter includes the Bayerischer Wald National Park, which became a biosphere reserve in 1981. High mountains (*Hochgebirge*), mostly 2,000 m or less in height, occupy the southern border with Austria. From west to east, the main groups are the Allgäu Alps (Madelegabel, 2,646 m), the Bavarian Alps (Zugspitze, 2,963 m), and the Berchtesgaden Alps (Watzmann, 2,714 m).

The middle mountains are largely forested, with marginal agriculture persisting but now largely dependent on government and European subsidies. These areas have experienced considerable emigration in recent decades and have weak economies. In spite of considerable potential for hydroelectricity, this has been little developed and energy supplies are often problematic. In some regions, notably the Black Forest, tourism is important, and is often seen as a focus for economic development. Yet the forested landscapes of all of these areas – a major attraction for tourism – have been considerably affected by acid precipitation, resulting in forest

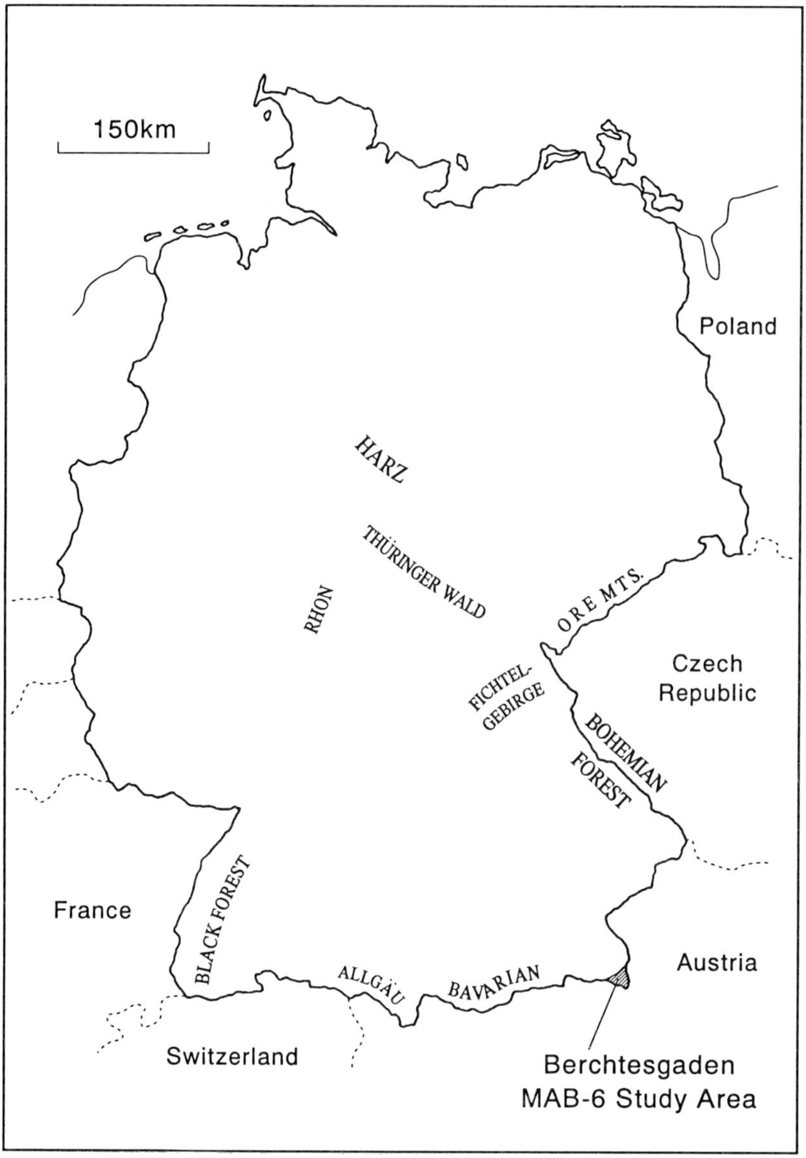

Figure 5.1 Germany: mountain ranges and the Berchtesgaden MAB-6 study area

decline (*Waldsterben*), which has negative effects not only on environmental stability but on all economic sectors.

The environmental and economic situation in the Alps is rather different. They are the pre-eminent tourist destination in Germany, for a variety of reasons. They are the only high mountains in the country – with considerable areas above the anthropogenic timberline and relatively well-preserved local cultures – and are easily accessible from major urban centres, especially Munich, which is close enough for day trips. Summer is the main season because of the low altitude of most of the resorts; less than a third of tourist nights are in winter. While the great majority of tourists are German, a few resorts have always attracted a wide clientele and the Berchtesgaden Alps, in particular, have recently seen an increase in visitors from the former communist countries of eastern Europe. However, total numbers of visitors to the Alps have changed little since the early 1980s, particularly because of overcrowding in peak seasons, limitations on development (regional planning began in the early 1970s) and increased ownership of second homes, many of which become retirement homes. Linked to these trends are increases in land and house prices and changes in demographic structure, with the emigration of young people and the immigration of older, wealthier people.

During this century, as tourism has increased its dominance and the transport infrastructure has improved, the agricultural sector has changed considerably. Part-time farming has become predominant in many areas, notably where holdings are small and tourism offers seasonal jobs; and while dairying has become important, especially in the Allgäu, extensive cattle grazing has also increased in recent years. One major causal factor has been subsidies which recognize the importance of a diverse mountain landscape for tourism and conservation. Forests are also used for cattle grazing which, together with large populations of deer, has had serious long-term effects on forest stability, exacerbated in recent years by the effects of acid precipitation. Since the late 1960s, considerable sums have been invested in reforestation and other investments to minimize natural hazards.

ORGANIZATION OF MAB-6 ACTIVITIES

The German National Committee for MAB was established in 1972 by the Ministry for Foreign Affairs. Until 1986, the federal Ministry for the Interior seconded the chairman and funded the secretariat; subsequently, the federal Ministry for Environment, Nature Conservation and Nuclear Safety has provided this support. The responsible minister appoints the members of the Committee. They include natural and social scientists from both government research institutions and universities, as well as representatives of the national UNESCO Commission, the German Research Association, and the federal Department of the Environment. There are also honorary members, some representing federal ministries and ministries of States with ongoing MAB research, others in their personal capacities. The chairman of the MAB

Committee appoints the secretariat, which consists of a secretary, another scientist, and two assistants.

The MAB Committee meets annually. Until 1991, all decisions concerning MAB activities were taken in plenary session. In 1991, the committee established four working groups – on biosphere reserves, land use and land-use changes, environmental education and ethics and environmental monitoring – which report to the annual meeting. The committee has been active in its support of international meetings, sponsoring fifteen from 1978 to 1990. These included four relating to MAB-6 (1978, 1981, 1984, 1985) and the first EuroMAB meeting (1987), as described in the introduction.

MAB-6 RESEARCH ACTIVITIES

MAB-6 research in Germany has comprised one major interdisciplinary project in the Berchtesgaden Alps of Bavaria (see Figure 5.1). The Garmisch-Partenkirchen region of the Bavarian Alps, where there is a long-established federal research station, was also considered. There were four reasons for the choice of Berchtesgaden: the existence of a range of ecosystems, from montane to alpine, with anthropogenic influences varying from minimal to substantial; the wealth of existing information; an existing infrastructure for field research; and the recent establishment (in 1978) of an Alpine Park, which includes both the National Park (20,780 ha) created in the same year and the 'forefield' to its north (25,820 ha; Figures 5.2 and 5.3). The Alpine Park and the MAB-6 test region are equivalent; this region was also declared a biosphere reserve in 1990. It ranges in altitude from 470 to 2,713 m.

Most of the National Park had been protected as a nature reserve since 1921; the higher altitudes had previously been a botanical reserve since 1910. The legislated objectives of the National Park emphasize the conservation of natural systems, research within them, and the dissemination of information. Much of the National Park consists of relatively untouched natural ecosystems, but there are a wide range of management issues:

(1) Forest management, including forest decline;

(2) The ecological and economic aspects of mountain agriculture (which is permitted in some zones of the Park);

(3) The impacts of summer tourism, especially relating to increased traffic, waste disposal, and cultural and economic aspects; and

(4) The conservation of the regional landscape.

The forefield includes a number of settlements (population *c*. 25,000) whose

Figure 5.2 Berchtesgaden MAB-6 study area, looking north over the Jenner (with hiking trail) to the forefield and the Untersberg in the distance

Figure 5.3 Berchtesgaden National Park, looking south to the Steinernes Meer (Funtensee test area)

economies are generally dominated by summer tourism, which began with the arrival of the railway in 1888. Tourism grew substantially after the Second World War, with stagnation in the 1980s and a recent increase associated with the end of the communist era in eastern Europe. About 80 per cent of the tourist income derives from the summer season. Both agriculture and forestry continue, but are no longer the basis for a viable economy.

A seminal event in the development of the MAB-6 project was a 1977 seminar on research in national parks, organized by the planning staff of the National Park and the Academy for Nature Protection and Landscape Management, established as a result of the 1973 Bavarian Law for Nature Protection. The project was further discussed at a seminar in Berchtesgaden, supported by UNESCO, in late 1978 (German National Committee, 1979).

In 1978, the federal Department of Environment contracted the Chair for Landscape Ecology at Munich-Weihenstephan Technical University to co-ordinate the preparation of the project. A questionnaire was sent to universities in Bavaria and other German states to establish their interest in taking part in the project. A total of 66 applications for participation were received. Subsequently, group meetings involving the various disciplines were held, in which the need for interdisciplinary research was debated and the initial stages of planning for co-ordination and the division of responsibilities were discussed. These meetings led to a feasibility study (Seibert, 1979).

In 1980 and 1981, the preliminary study was reviewed by the federal Interior Ministry, the federal Department of Environment, the Bavarian State Ministry for Environment and Planning, the National Park Administration and its authors. They concluded that the proposed multidisciplinary approach would not lead to the level of synthesis important for interdisciplinary ecosystem research, as had been envisaged when the study was requested. Thus, it was decided that constructive criticism of the preliminary study was required. This was done by two private firms specializing in system analysis – who came to similar conclusions.

In 1981, one of the firms (ESRI-Munich) was asked to work with the Chair for Landscape Ecology at Munich-Weihenstephan Technical University and the Administration of the National Park to develop a new feasibility study. This led to a complete reorganization of the philosophy of the project. In particular, the importance of integrative approaches to research (i.e. interdisciplinarity, rather than multidisciplinarity) was given far greater stress than before, and the availability of existing information suitable for model development was recognized as a key issue in defining the relative importance of individual research activities (d'Oleire-Oltmanns, 1985). The project was to be problem-orientated, with an emphasis on the development of applied, interdisciplinary methodologies. This approach drew particularly on the Munich-Weihenstephan group's experience of a previous attempt at integrated research (on intensive agriculture in the Ingolstadt region of Bavaria: Bachhuber et al., 1983), which had not been successful because of inadequate co-ordination. In consequence, it was realized that such work required strong co-ordination, to ensure that research took place within a well-defined framework, and that central, open data management was essential.

The project was divided into four phases over 11 years (1981 – 1991). The first phase, only a few months long, was the preliminary phase in which the plausibility of planned approaches was evaluated through research in a small portion of the project area (the Jenner) characterized by a high density of data from various disciplines. The second phase, in 1982, involved the establishment of a management framework by ESRI-Munich, and the evaluation, processing, and storage of available information in the GIS developed throughout the project. It also included the first model runs, in order to establish where essential data were lacking and to identify possibilities for developments in system theory. The studies begun on the Jenner in 1981 were extended to include initial simulation modelling.

Two reports summarize this initial phase, whose critical objective was to identify the optimal data and methods required for answering specific applied questions. The first report covers aims, problem formulations and methods (German National Committee, 1983a). The second describes scenarios and their evaluation for the Jenner test area. The scenarios consider forest decline; alpine pasture (*Alm*) management, in particular the influence of human activities on vegetation and insects and the socio-economic implications of changed management; and summer tourism, in particular its effects on alpine ecosystems and their zoofauna and aesthetic – psychological research for the assessment of changes in use. A final chapter describes the sources of data and their processing and evaluation (German National Committee, 1983b).

The principal phase of research began in 1983 and continued until 1988. It was co-ordinated by four institutions: the Chair for Landscape Ecology at Munich-Weihenstephan Technical University, the Administration of the National Park, ESRI-Munich and Planning Office Dr Schaller (in Kranzberg, near Munich). The involvement of private companies was critical for the implementation of such a long-term project. In Germany, research scientists at universities can only be supported by federal funds for non-renewable 5-year periods. The incorporation of the private companies permitted a small co-ordinating team to be established and continue throughout the life of the project. Most of the members of this team started their work at Munich-Weihenstephan and subsequently worked with the private companies, returning to the university as lecturers to pass on their experience.

A total of seventy-eight scientists and planners from institutions in Germany and other European countries were involved. Participants were chosen according to their willingness to participate in an interdisciplinary, problem-orientated dialogue and undertake projects within the overall research framework. Funding came from the federal Ministry for Research and Technology; the federal Ministry for the Environment, Nature Conservation and Reactor Safety; the federal Department of Environment; and the Bavarian State Ministry for Environment and Planning. Three major objectives were stated (Ashdown et al., 1988):

(1) To solve current ecological problems concerning land use in high mountain ecosystems, using data analysis to avoid problems associated with exceeding carrying capacity;

(2) To validate the data and results in real situations in order to improve the plausibility and transferability of the methods; and

(3) To ensure transferability of the techniques to solve problems in other high mountain ecosystems.

The project as a whole was based on geographical theories of landscape ecology (cf. Haber, 1990) and the systems approach developed in the Swiss MAB-6 programme and further elaborated during the Berchtesgaden project (Figures 5.4 and 5.5). The best source for detailed information concerning these conceptual bases is the two-volume summary report produced in 1991 at the end of the 4-year synthesis phase (1988–91) by the same small group that undertook the initial development of the project (Kerner et al., 1991). This report also lists all publications produced during the course of the project. Similarly, the development of the Berchtesgaden GIS – an integral part of the project – will not be described here: the best source is Ashdown and Schaller (1990) (Figure 5.6). Nevertheless, one should recognize that these methodological products were among the most important results of the project.

Research within the project considered both the whole project area (National Park and forefield) and, in greater detail, four test areas covering about 20 per cent of the area, each with a range of different land-use types (Figure 5.7). These were:

(1) Jenner (3,292 ha; 603–2,350 m): a montane–alpine area on the boundary between the National Park and the forefield, with a substantial tourist pressure, where impacts of summer and winter tourism, as well as alpine and forest grazing and agricultural problems, were studied;

(2) Funtensee (470 ha; 1,601–2,158 m): a high alpine area in the National Park dominated by relatively unmodified abandoned pastures, where the effects of low-impact tourism were studied;

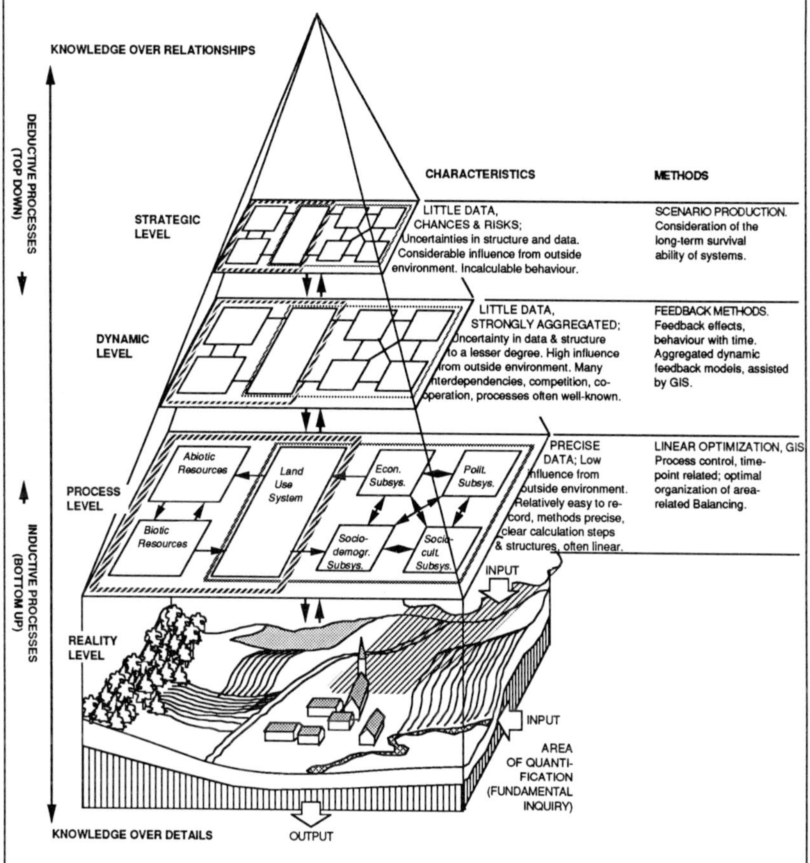

(adapted and modified from GROSSMAN, 1983)

Figure 5.4 Graphical representation of the hierarchical systems method as applied to an ecosystem complex, showing the sub-systems at three levels of complexity (from Ashdown and Schaller, 1990, with permission)

(3) Ramsau (2,671 ha; 670–2,607 m): an entire valley system centred on the tourist village of Ramsau, on the boundary between the National Park and the forefield, with many farms, where research concentrated particularly on socio-economic issues; and

(4) Untersberg (1,228 m; 719–1,972 m): a largely unmodified catchment in a massif on the northern edge of the forefield, with many relatively

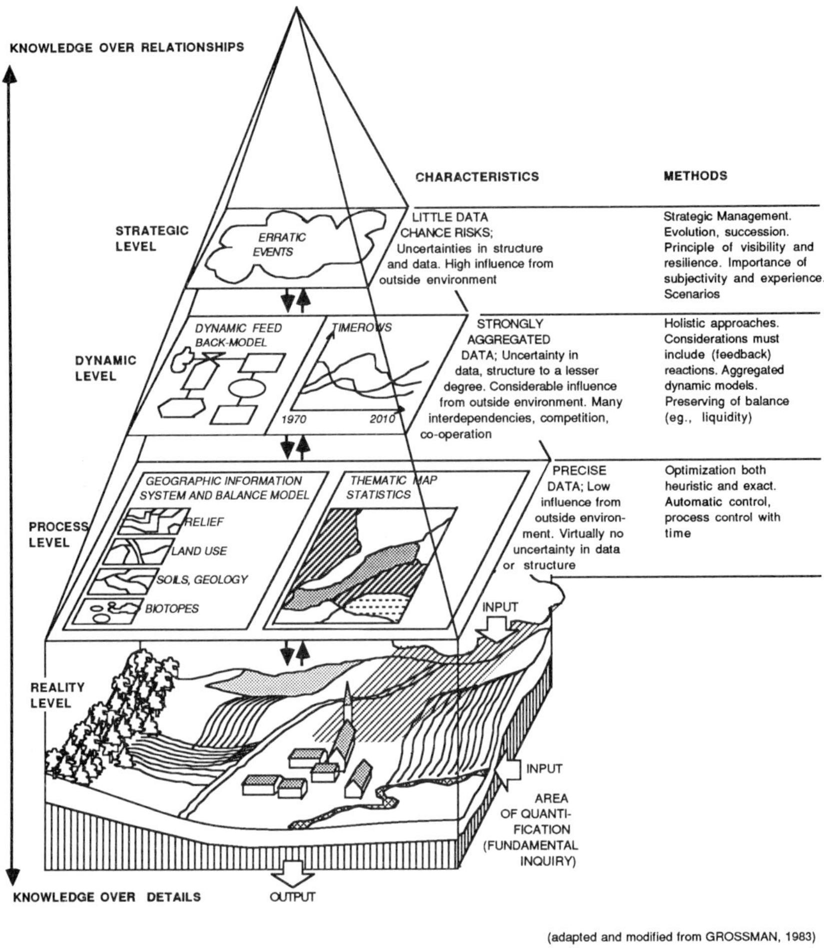

Figure 5.5 Graphical representation of the hierarchical systems method as applied to an ecosystem complex, showing the different models applied at three levels of complexity (from Ashdown and Schaller, 1990, with permission)

pristine montane forests and large, unmanaged former pastures (i.e. declining alpine grazing), where studies of forests and abandoned pastures were made.

In addition to the research conducted according to the overall plan for the

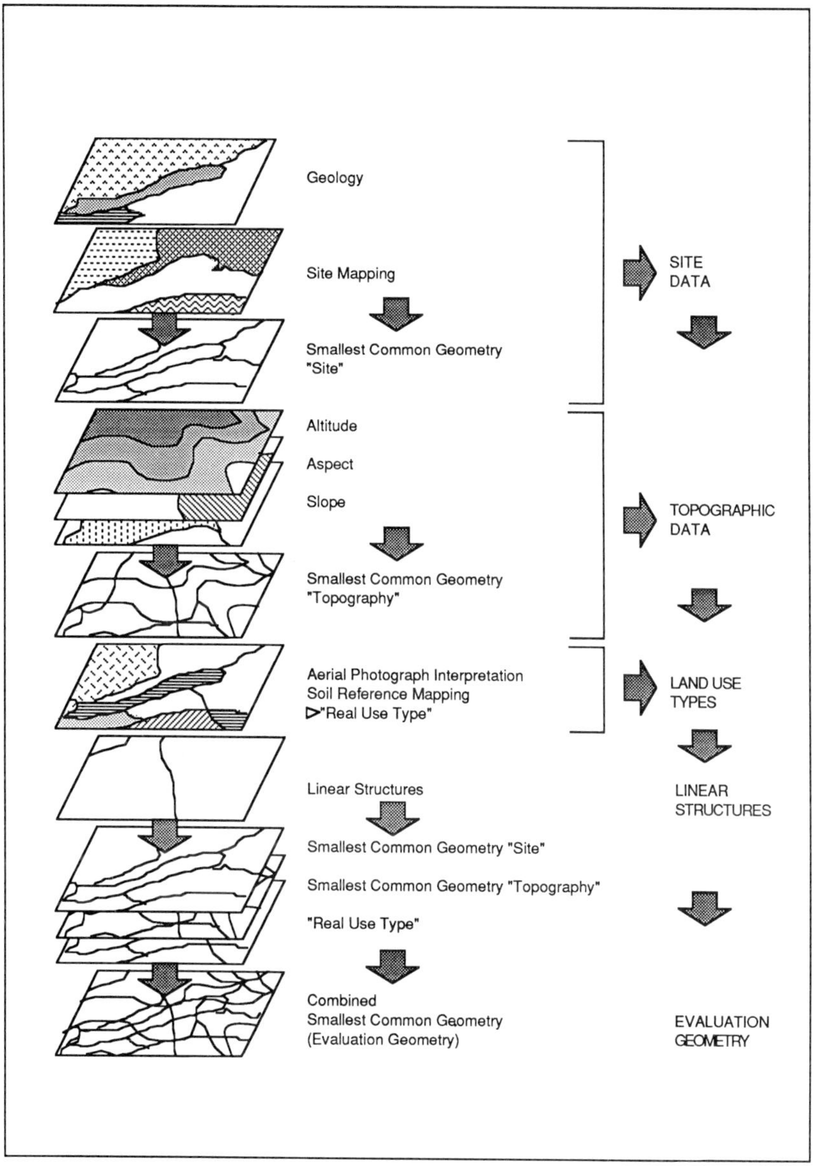

Figure 5.6 The basic structure of the Berchtesgaden geographic information system (GIS) (from Ashdown and Schaller, 1990, with permission)

principal phase, three special studies were conducted. The purpose of these studies, as with the scenario-based studies of the initial phase, was to test the methods in their current state of evolution, using currently available

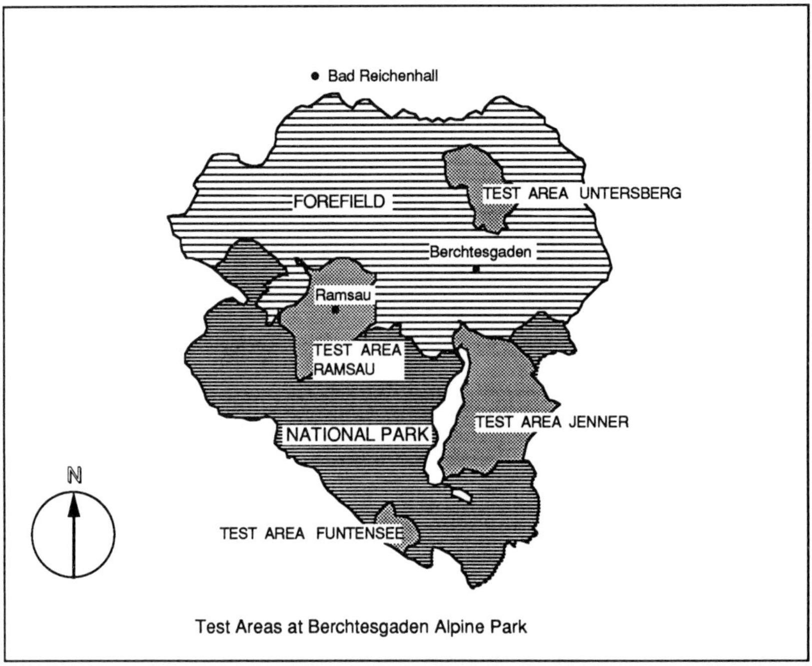

Figure 5.7 The Berchtesgaden study area, showing the four test areas (from Ashdown and Schaller, 1990, with permission)

data. The first of these studies considered the impacts of summer tourism on vegetation, extending the study on the Jenner in the initial phase, and was completed in 1987. The study used the information in the GIS together with field research to evaluate the critical factors determining the sensitivity of different areas to trampling. This predictive model was then tested in the Funtensee area and subsequently extrapolated to the entire National Park. The resulting maps and other knowledge were used to develop proposals for monitoring and management (Spandau, 1988).

The second special study was an opportunity to apply the methodologies to a real situation: the possible environmental and socio-economic effects of a Winter Olympic Games in Berchtesgaden, as proposed by the German National Olympic Committee in 1983 (Haber, 1986). This was a major topic of debate within the region. The study was undertaken in 1985–86, and would have continued longer, but the International Olympic Committee rejected Berchtesgaden's bid for the Games in the first round. Four variants of development were considered through scenario building:

(1) No Olympic Games: a null hypothesis for comparison;

(2) 'Soft' Olympic Games: no new ski runs, maximum use of existing facilities;

(3) Wider planning: new ski runs, but with development integrated into the wider region; and

(4) 'Hard' Olympic Games: unrestricted development on the Jenner, making Berchtesgaden a leading winter sports centre.

Figure 5.8 compares some of the projected effects of the 'soft' and 'hard' scenarios. The study involved not only the project scientists, but also local people, who participated in multi-day workshops at which the study was explained, information sought and, later, results presented. The project team was able to use the integrated methodological approaches to provide an objective analysis of the potential positive and negative impacts of the Olympic Games which was widely disseminated and utilized. The conclusions of the evaluation were that, in general, the effect of the Olympic Games would be neither especially negative nor positive. Revitalization of the region was necessary, but the Games would have only a short-term benefit. Long-term planning and the promotion of the existing attractions of the Park – its environment, culture and individuality – were the best approaches. This meant an emphasis towards small, rather than large, hotels, which not only were more desirable to tourists, but also could provide more jobs for local people. In addition, the critical limitations of the traffic infrastructure were underlined.

A later special study (1989–90) also used the approach of integrated research, scenario construction and workshops to address the question of the long-term future of agriculture in the region. However, this did not have as great an impact, partly because the results were not disseminated and discussed at a workshop because of decreasing funding in the synthesis phase; but also because it was recognized that external influences, such as international agricultural and economic policies, were critical factors in determining future patterns (Kerner and Spandau, 1990).

The final activity of the MAB-6 project was a seminar in 1991, at which the final summary report (Kerner et al., 1991) was presented and discussed. This report consists of two volumes, with four sections: objectives and issues of the project; concept of contents and methods; project activities; critical assessment and recommendations. The first three sections include both methodologies and summaries of the results of research and case studies. Two additional major methodological publications derived from the project. One was a report on landscape assessment, based on an environmental psychological approach to landscape aesthetics (Nohl and Neumann, 1986). This was prepared because of the requirements in the federal nature protection law for objective criteria for landscape impact

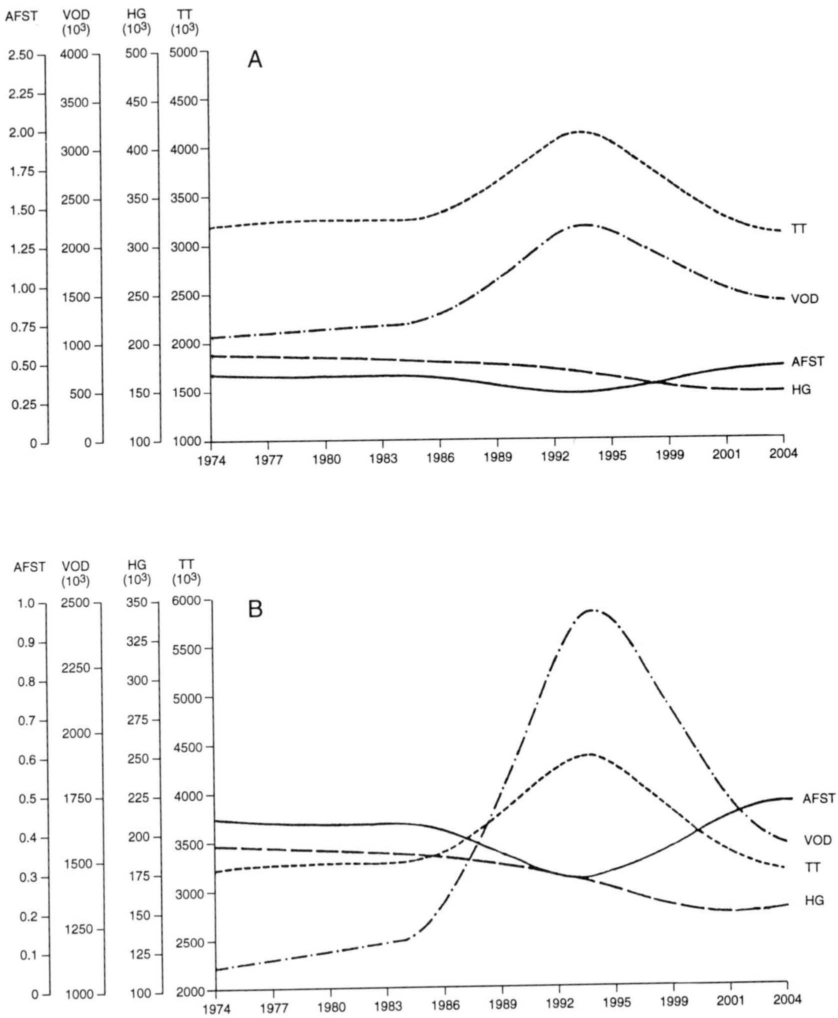

Figure 5.8 Comparison of the results of the 'soft' (A) and 'hard' (B) scenarios for the development of the Winter Olympic Games in Berchtesgaden. TT = Tourism Total, HG = Hotel guests, VOD = Visitors in one day, AFST = Attractiveness for summer tourism (redrawn from Haber, 1986, with permission)

assessment as part of the planning process. The other is a book on the conceptual bases for applied ecosystem research, prepared by one of the members of the co-ordinating group (Tobias, 1991).

Another important publication resulting from the project is the final report of the specialist disciplines (Haber et al., 1990). This begins with brief introductions to the conceptual bases for the project, the project region and the test areas. Subsequently, the results of work by each specialist research group – publicity, botany, limnology, zoology, agriculture, geology/soils, climatology, air pollution, hydrology, technology (land use and settlement), forestry, socioeconomics, tourism, economics, demography, landscape aesthetics and traffic – are presented and discussed. The report concludes with a summary of the special studies on the Winter Olympic Games and summer tourism and a complete bibliography of the project.

CONCLUSIONS

The Berchtesgaden project has had a number of significant results both for the project area and at wider scales. The GIS developed during the project has proved a functional tool for research, training and management particularly in the National Park and, to a lesser extent, in the forefield. In addition, the integrated systems-based scenario approaches were valuable in providing evaluations of current and future patterns of forest damage, agriculture and the ecological impacts of recreation. Perhaps the greatest test and success of these methods was their use in the objective assessment of the future of the region in relation to the proposed Winter Olympic Games which, like the special study on the future of agriculture, brought many local people into the project.

Nevertheless, the full potential of the integration of the GIS, scenario-building and public workshops has not yet been realized through the development of an integrated management approach for the biosphere reserve. The Bavarian State Ministry for Environment and Planning has recently requested that a master plan for the National Park be developed over the next 2 years. This process will begin by updating the land-use component of the GIS, based on the interpretation of recent aerial photographs (the current information is based on 1980 photographs). Subsequently, the updated GIS will be used as a basis for the development of the master plan, using a methodology developed, but not applied, during the MAB-6 project. This work will be undertaken by a private company in consultation with the Ministry and the National Park Administration. A greater test of the methods developed during this MAB-6 project would be the development of scenarios and plans for the entire biosphere reserve, which includes a wide range of potential resource conflicts. However, this is not planned.

The experience gained in Berchtesgaden is being increasingly widely applied in biosphere reserves both in Germany and abroad. One example is the current development of guidelines for the protection, management and (sustainable) development of German biosphere reserves by the national MAB Committee's permanent working group on biosphere reserves. Another example is the development of a general methodology for ecological monitoring in German biosphere reserves, which is being tested in Berchtesgaden, with support from the federal Department of the Environment. This work is important, as it will require the implementation of the time dimension in the GIS approach; so far, consideration of this dimension has not been very well developed. Thus, the methods of data collection and storage and interdisciplinary interaction developed in Berchtesgaden are increasingly being used for monitoring and management in present or eventual future biosphere reserves, not only those in mountains – including Changbaishan (China) and Mount Carmel (Israel) – but also in other biomes, such as the Wattenmeer.

Another project under development concerns the Bohemian Forest, which includes the Bayerischer Wald National Park and biosphere reserve in Germany, the Šumava biosphere reserve in the Czech Republic and the Muhlviertel region of Austria. This tri-national MAB pilot project will involve the governments of the respective regions (Bavaria, Czech Republic, Upper Austria). It is being co-ordinated by Planning Office Dr Schaller, and has been partially funded by the Allianz Stiftung (Foundation) of Munich. In summary, while MAB-6 activities *per se* have concluded in Germany, the consequences of past activities have been, and continue to be, significant not only for the Berchtesgaden area, but also for many other places in both Germany and other countries.

CHAPTER 6

SPAIN

MOUNTAIN REGIONS

While somewhat more than half of Spain is occupied by the flat mesetas of Old and New Castile, most of the rest of the country is mountainous (Figure 6.1). The highest mountain ranges are in the extreme south and north of the country. The country's highest summit is Cerro de Mulhacén

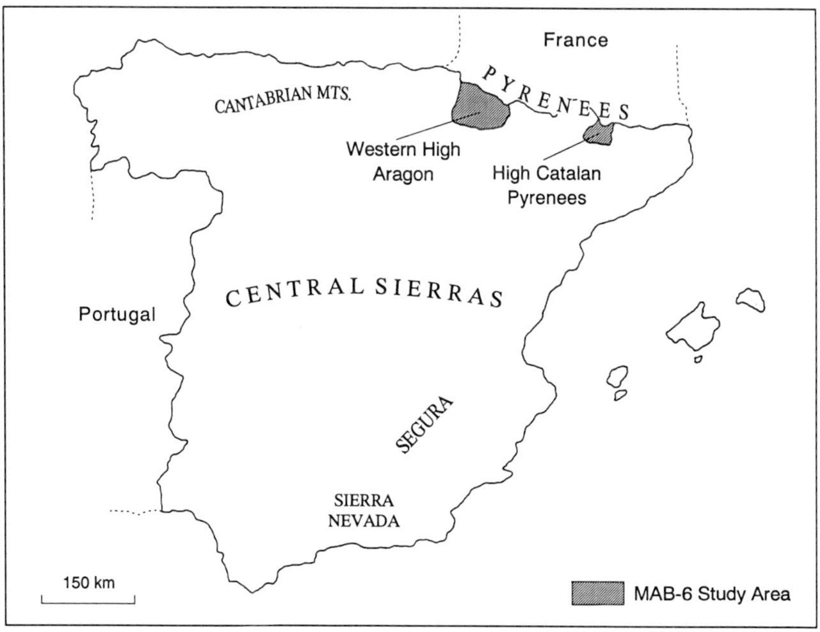

Figure 6.1 Spain: mountain ranges and the MAB-6 study areas

(3,482 m) in the small Sierra Nevada, south-east of Granada. In contrast, the second highest peak, the Pico de Aneto (3,404 m) is in the Pyrenees. These have more than 50 summits over 3,000 m, and stretch across the northern border with Andorra and France, almost from the Bay of Biscay to the Mediterranean. The other major northern range is the Cantabrian Mountains, south of the Bay of Biscay, whose highest peak is the Peña Vieja (2,615 m) in the Picos de Europa.

South of the Cantabrian Mountains is the meseta of Old Castile. This is separated from the meseta of New Castile by the Central Sierras, which stretch across the middle of the country from the Portuguese frontier almost to the Gulf of Valencia. From west to east, they include the Sierra de Gredos (Pico de Almanzor, 2,592 m), the Sierra de Guadarrama (Pico Peñalara, 2,469 m) north of Madrid, the Soria-Teruel group (Moncayo, 2,316 m) and the Sierras de Javalambre and Gúdar (Javalambre and Peñarroya, both 2,020 m). The Sierra de Segura (La Sagra, 2,381 m) is in Murcia, north of the Sierra Nevada. There are also many other ranges whose highest summits have an altitude of less than 2,000 m.

In general, the mountains of Spain may be described as marginal areas which have been greatly affected by external socio-economic influences during the twentieth century. Given their great diversity in terms of climate, accessibility and natural resources, trends can only be generalized; there are exceptions to all of the points made below. In most areas, the highest population densities were around the turn of this century. Since then, the mountains have experienced a decrease in population – some have only one-quarter of their population in 1900 – and populations are generally ageing. These trends have only been reversed in a few areas which are either close to major population centres, such as Madrid, or have become centres for the development of tourism, for instance around ski resorts in the Pyrenees. However, all tourism is seasonal, generally relying on a winter season of no more than 4 months and a summer season of no more than 3 months. There has been great development of second homes and self-catering accommodation which are empty for most of the year.

Many villages have been abandoned, for a variety of reasons. These include the unprofitability of traditional agricultural economies, the draw of more attractive employment opportunities elsewhere and the physical loss of villages and surrounding land drowned by reservoirs or within the areas of the major reforestation projects that began in the 1940s. Reservoirs have been built since the early years of this century to regulate water supplies to the plains and provide electricity. Reforestation projects have had many purposes: to create employment; to minimize soil erosion, especially in southern and Mediterranean areas; to provide a new resource base, especially in abandoned areas; or to provide raw materials for the pulp and paper industries. Reforestation projects have generally become less important since the mid-1970s, as fewer high-priority areas were left and communities become less willing to lose communal land, usually used for grazing.

The maintenance of this balance between summer pastures and valley resources – pastures and meadows to provide hay for cattle which are kept in stalls for much of the winter – is critical to the survival of cattle-raising for beef or milk, for which the demands continue to grow in an increasingly urbanized country. These demands have intensified pressures for cattle-raising; but this is only profitable in large operations which use modern techniques and have adequate access to winter fodder and markets. Labour-intensive activities, such as the raising of sheep and goats and the cultivation of cereals, are generally declining; agriculture now tends to be concentrated on the most fertile, easily-accessible land where mechanization is possible. The principal exceptions to these rules are where high-value goods – such as ham, fruit, honey, or cheese, depending on the area – can be produced and marketed.

To some extent, the economic situation of the remaining mountain inhabitants, who receive subsidies from both the Spanish government and the European Community, has improved in recent decades. Nevertheless, mountain communities are increasingly dependent on external processes and decisions, which affect not only the demand for the goods they produce, but also the price of those they need to survive as the old self-sufficient system declines. Throughout the mountains of Spain, physical infrastructure, such as terrace walls and irrigation channels, is falling into disrepair, and abandoned fields and terraces are being recolonized by shrubs and trees; the changes in the landscape, that began in the early years of this century and have intensified in recent decades, continue.

ORGANIZATION OF MAB-6 ACTIVITIES

In 1970, after the 1968 UNESCO Conference on the Biosphere, the Spanish National Committee for UNESCO – for which the Ministries for Education and Science and of External Affairs are jointly responsible – established a National MAB Committee and appointed its first chairman. There have been only three chairmen during the period of the MAB programme: the first from 1970 to 1978, the second from 1978 to 1990 and the third since then. The committee includes scientists and representatives of regional governments, ministries and society in general. Initially, the committee was composed principally of scientists from various disciplines; in recent years, the proportion of representatives of regional governments and ministries has been increasing. There are now four vice-chairmen, responsible for research, international relations, education and finances. Within the committee, a smaller permanent commission follows the course of project proposals and approved projects in greater detail. However, fundamental decisions regarding all projects are made in plenary session.

The secretariat of the committee is appointed by the chairman, and currently consists of an executive secretary and two functionaries. Initially, financial support for the secretariat came principally from the National Commission for UNESCO. However, in recent years, more support has come from the Direction for Environment in the Ministry of Public Works and Urbanization (MOPU), where the executive secretary is based, and regional governments.

MAB-6 RESEARCH ACTIVITIES

There have been two groups of MAB-6 research projects in Spain, both in the Pyrenees and Pre-Pyrenees to their south (see Figure 6.1). The first began in the 1970s, in western High Aragon (*Alto Aragon occidental*; central Pyrenees), and included three projects: a multidisciplinary, integrated study of the resources of western High Aragon, an integrated study of Pyrenean pine forests; and an integrated study of Pyrenean ecosystems with large herbivores. A later project was undertaken in the 1980s in the High Catalan Pyrenees (*Alt Pirineu*: eastern Pyrenees).

Western High Aragon

Western High Aragon has a transitional Mediterranean-continental climate, with an Atlantic influence that decreases sharply from its northern boundary. Thus, there are sharp altitudinal gradients in both climate and ecosystems. The area has been inhabited since pre-Roman times by nomadic and then transhumant and settled people. There were major increases in populations of both people and grazing animals in the thirteenth century (followed by major decreases in the fourteenth century) and from the fifteenth to nineteenth centuries. The stimuli to the latter period of growth were, first, stock-breeding (cattle and sheep) and, particularly after 1850, the cultivation of potatoes (Figure 6.2).

The systems that evolved from the mid-nineteenth century, and persisted until the 1960s, involved slash-and-burn cultivation and the grazing of sheep and goats on both agricultural land and communal forests; timberline was lowered as summer pastures expanded (Figure 6.3). The period of highest population density was around the end of the nineteenth century, when considerable emigration began (Figure 6.4). After the First World War, villages with less-favourable locations began to be abandoned; a trend which intensified from the 1940s, after the Spanish Civil War, as cattle-

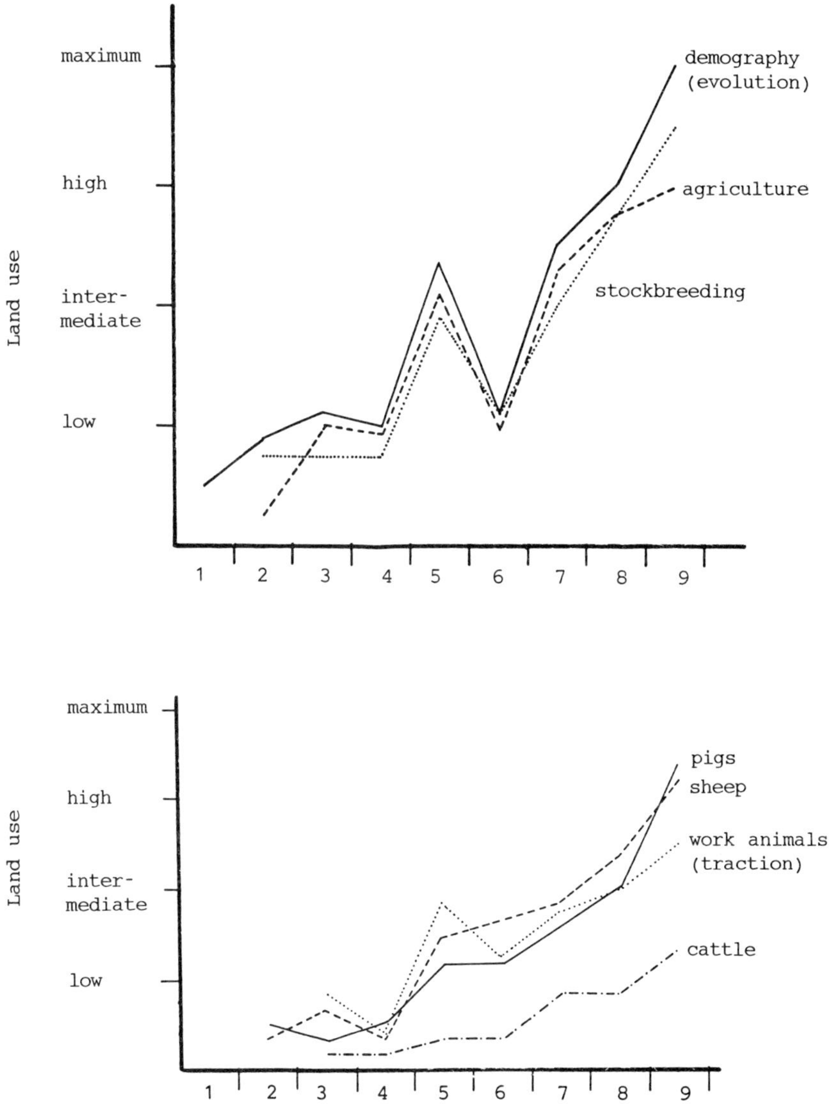

Figure 6.2 Demographic and agricultural trends in the major prehistoric and historical periods in the High Aragon Pyrenees. 1 = Palaeolithic; 2 = Neolithic; 3 = Early documented history; 4 = fourth and fifth centuries; 5 = Expansion in the thirteenth century; 6 = fourteenth century; 7 = Expansion between the fifteenth and eighteenth centuries; 8 = Agricultural revolution of the eighteenth century; 9 = End of the nineteenth century (from Chocarro et al., 1990, with permission)

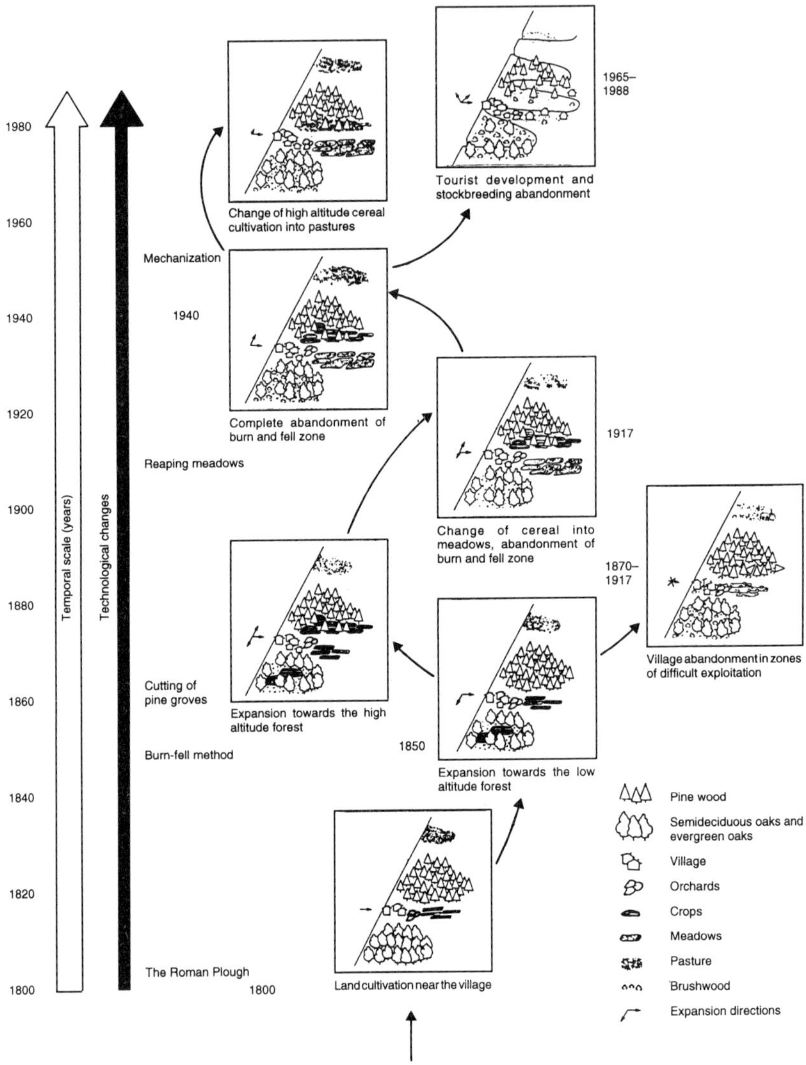

Figure 6.3 The sequential use of natural resources in the Pyrenean valleys of High Aragon, 1800 to present (from Chocarro et al., 1990, with permission)

breeding increasingly became the backbone of the economy (Figure 6.5). At present, the remaining villages tend to rely on either seasonal tourism or cattle-breeding. Much high-quality agricultural land has been lost through the construction of both tourist accommodation and dams for hydroelectricity (Chocarro et al., 1990; García-Ruiz and Lasanta-Martinez, 1990; Figure 6.6).

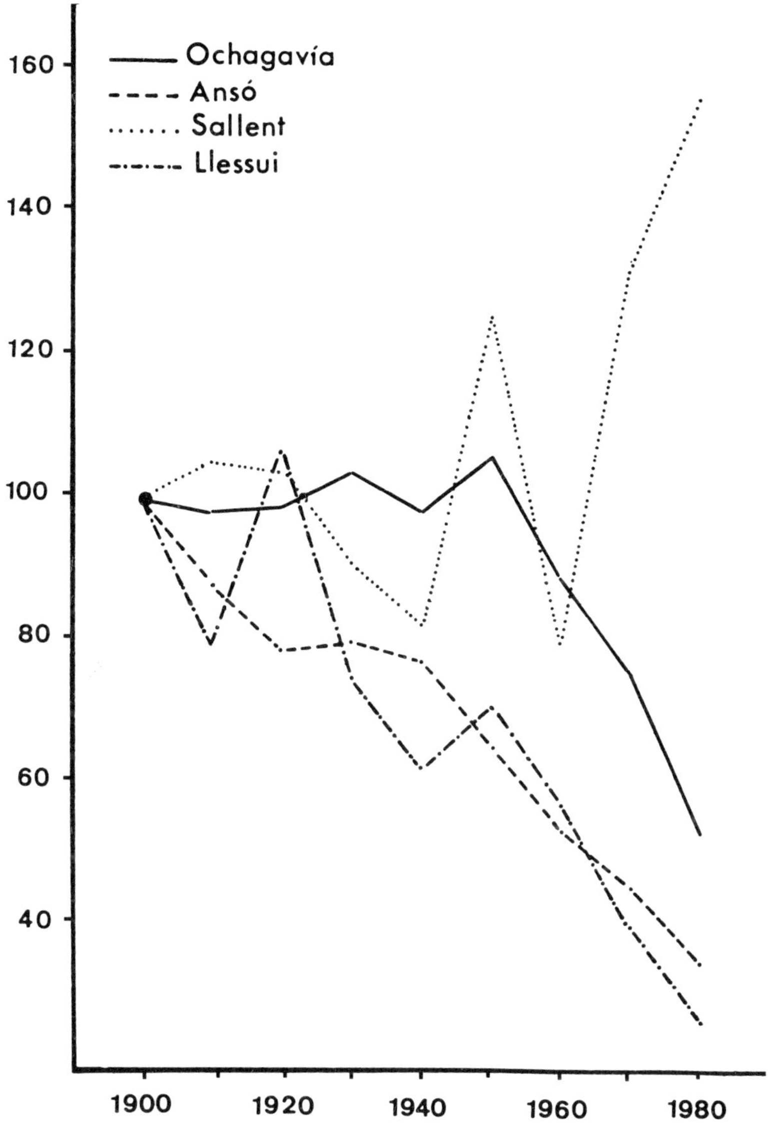

Figure 6.4 Percentage change in the population of four villages in High Aragon, 1900 to 1980 (from García-Ruiz and Lasanta-Martinez, 1990, with permission)

The MAB-6 projects in western High Aragon have been co-ordinated from a research institute in Jaca, Aragon. Initially, this was the Pyrenean Centre for Experimental Biology (CPBE), established in 1963 by the Higher Council for Scientific Investigation (*Consejo Superior de Investigaciones*

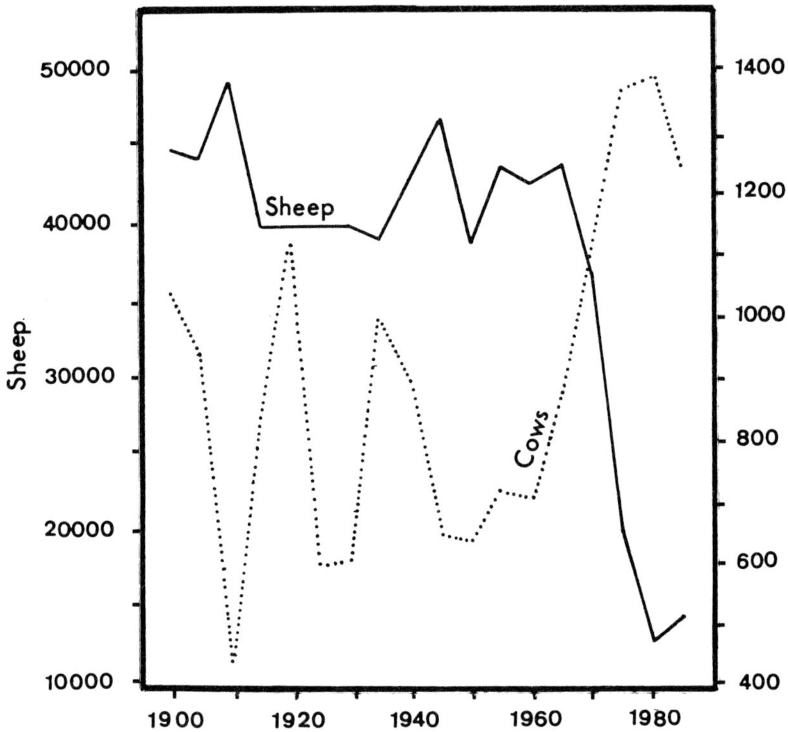

Figure 6.5 Changes in the number of livestock in the Ansó Valley, 1900 to 1982 (from García-Ruiz and Lasanta-Martinez, 1990, with permission)

Cientificas: CSIC), the national agency for the support of scientific research, which maintains research institutions throughout Spain. In 1983, the CPBE was merged with the Institute of Pyrenean Studies to form the Pyrenean Institute of Ecology (*Instituto Pirenaico de Ecologia*: IPE). The Institute of Pyrenean Studies had been established by the CSIC in 1942, as an instrument for co-ordinating research in the Pyrenees by scientists from all natural and social science disciplines. The means by which this was done included the publication of the journal *Pirineos* (from 1945) and monographs (from 1943), as well as the organization of scentific meetings, particularly the quadriennial International Congresses of Pyrenean Studies, which were held alternately in Spain and France from 1950 to 1974. In addition, a library was established at Zaragoza, where field equipment was also maintained. From the establishment of the CPBE in 1963, it had the same director as the Institute of Pyrenean Studies, who brought the library to Jaca in 1968.

Figure 6.6 The almost deserted village of El Pueyo de Jaca, western High Aragon; most of the fields have been drowned by a hydroelectric reservoir behind a dam on the Rio Gállego

The IPE continues to publish *Pirineos*, which has included papers on research outside the Pyrenees since 1985; the journal of the CPBE (*Publicaciones del Centro Pirenaico de Biologia Experimental*) ceased publication when the two institutions were merged. These two publications are the primary sources of published material concerning the MAB-6 projects in western High Aragon. The only significant English-language sources of information on these projects are the guide for the 1986 field trip of the Commission on Mountain Geoecology of the International Geographical Union (Puigdefábregas et al., 1986) and a special issue of *Mountain Research and Development* (Puigdefábregas et al., 1990).

The first two MAB-6 projects effectively began in 1971, continuing work begun at the CPBE as part of ICSU's IBP. The multidisciplinary project on the western High Aragon ran until 1984 and the project on pine forests until 1982. The other project, on ecosystems with large herbivores, started in 1976 and effectively finished in 1990, although work on its themes has continued. All of these projects were supported through the usual system of funding from the CSIC to the CPBE and, later, the IPE. Most of the work was undertaken by the scientists of the CPBE/IPE, often working with graduate students, who presented their results as theses and dissertations. Scientists from other CSIC institutes also participated in some of the research. Most of the references cited below are major monographs and dissertations. However, much of the most important work has been published in papers in journals and some, particularly dissertations, remains unpublished. There have been no final synthesis reports.

The multidisciplinary study of the resources of western High Aragon considered the whole of the region (*c.* 3,750 km^2), at altitudes from 400 to 2,800 m. It was an overall study of the region, by scientists from a wide range of natural and social scientific disciplines. Its objective was to provide a comprehensive description of the region's biophysical resources and human history and ecology, as a basis for both education and planning. There was no overall framework of research; the general concept was to prepare a series of monographs that would eventually be synthesized into an overall descriptive monograph.

As well as more than 100 scientific papers, major publications deriving from the programme concern geology (Puigdefábregas, 1975; Soler-Sampere and Puigdefábregas, 1972), climatology (Creus, 1983; García-Ruiz et al., 1986), botany (Balcells, 1976; Villar, 1980, 1982), zoology (Martinez-Rica, 1979a,b; Palanca, 1987; Pedrocchi, 1987) and human ecology and land use in both historical and current perspectives (Balcells, 1976, 1984; García-Ruiz, 1976). Many of the latter reports implicated the results of the other two projects, described below, and another report (Anglada et al., 1980) drew on the experiences of the project to propose policies for the management of mountain resources, in relation to the law for mountain agriculture that was currently in preparation and passed in 1982. Consideration has been given to the preparation of a final synthesis and an exhibition of the overall results, but neither has been implemented.

The project on the pine forests focused on western High Aragon, in 60,000 ha of forests at altitudes from 600 to 2,000 m, with particular emphasis on two forests south-west of Jaca: El Boalar (950 m) and San Juan de la Peña (1,230 m). Fieldwork finished by 1982 and was briefly summarized, with a bibliography that also includes studies from the previous project, by Balcells (1983). Most of the major results of the research have only been compiled as unpublished doctoral dissertations; over forty papers also resulted, but much of the work has never been published.

There were five themes of research. The first was the preparation of a typology of the pine forests, according to species, primary production, climate regime and forest structure (Puigdefábregas, 1981). The second theme considered human impacts on the pine forests of High Aragon, in three phases: from prehistoric times to the eighteenth century, when the wood was used locally; the period from the eighteenth century, with increasing levels of export; and the current period, emphasizing the effects of forestry policy in this century. The third theme focused on primary production and biogeochemical cycling in typical forests, with research on the demography of twenty-three pine forests, the distribution of organic matter (at three sites), nutrient distribution and cycling (at El Boalar and San Juan de la Peña) (Alvera, 1990) and studies of foliage. The fourth theme considered the influences of primary and secondary consumers at San Juan de la Peña, with research on small mammals, boars, birds, epigean invertebrates (Pedrocchi et al., 1985) and lepidoptera, which are frequent defoliators. The final theme concerned processes of decomposition (Franch, 1985).

The project on ecosystems which are utilized seasonally by large herbivores considered an area of *c.* 17,000 km^2 across the Pyrenees, at altitudes from 400 to 3,150 m. Specific study sites included the Puertos de Goriz (4,800 ha) of Ordesa and Monte Perdido National Park and the Puerto de Aisa (1,300 ha, 1,600–2,650 m) in the headwaters of the Aisa valley. Both transhumant livestock (cattle, goats, horses and sheep) and wild herbivores (mainly chamois and grasshoppers) were considered. The overall objective of the project was to develop models of the traditional modes of exploitation and their evolution in response to external influences, particularly in relation to current and accelerating processes of socio-economic change, to project the future for the study areas.

The project included studies of the influences of grazing animals on production systems with different botanical and topographic characteristics; interactions between consumers (Figure 6.7); relationships between abiotic and biotic factors with regard to the exploitation of high mountain ecosystems by livestock and wild herbivores; anatomic and ecophysiological studies of large herbivores; catalogues of parasites; and studies of characteristic ecological complexes. In later years, the research increasingly concentrated on the valley pastures near villages, which are an essential component of the grazing economy. These pastures experience strong anthropogenic influences (irrigation, fertilization, grazing).

The majority of the research undertaken within this project was prepared as doctoral dissertations (e.g. Chocarro, 1990; Fillat, 1980; García-González, 1987; Gómez, 1989) or published in various journals (mainly Spanish) and proceedings of conferences, particularly those of the Spanish Society for the Study of Pastures (*Sociedad Española para el Estudio de los Pastos*). While no synthesis has been made, significant parts of the research have been summarized in papers (e.g. Chocarro et al., 1987, 1990; García-González et al., 1990; Montserrat and Fillat, 1990). The project effectively ended in the early 1990s, when research on its themes received financial support from various national sources and through programmes supported by the European Communities. Thus, the work continues in the Pyrenees of High Aragon, particularly in the Asia and Broto valleys (partly in the Ordesa and Monte Perdido National Park), often in collaboration with French scientists.

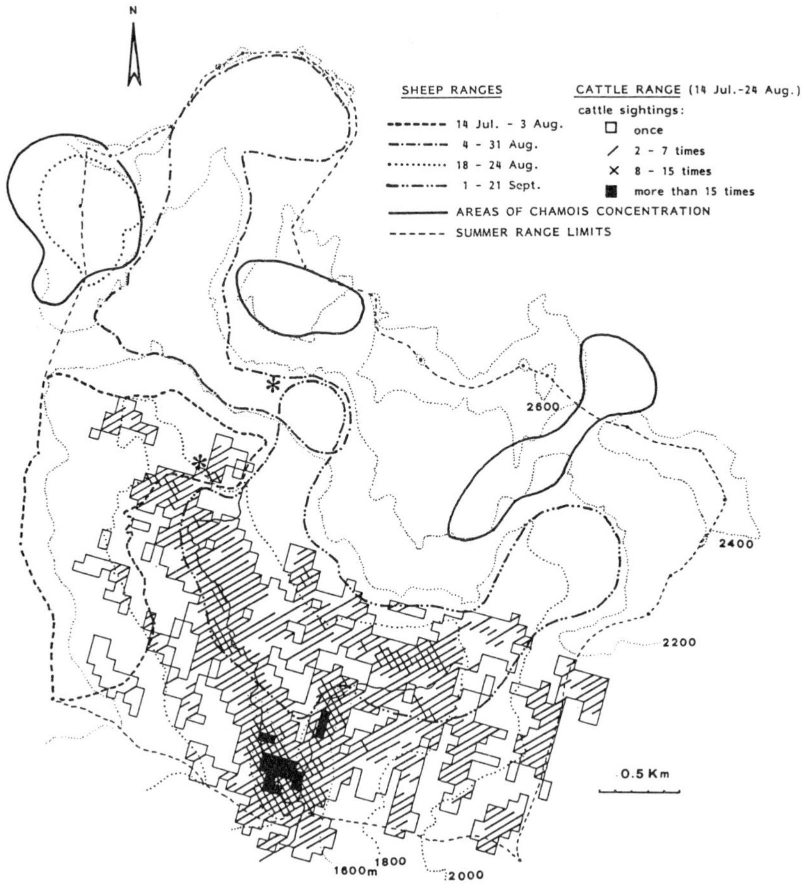

Figure 6.7 Spatial distribution of ruminants in the Aisa summer range in 1986. Sheep ranges indicate the customary courses used by a herded flock for a specific period. Cattle presence is expressed as the number of sightings (one or more animals) per quarter-hectare unit. Cattle are free-ranging and chamois are wild. The night-resting sites of sheep are shown by asterisks (from García-González *et al.*, 1990, with permission)

High Catalan Pyrenees

The High Catalan (*Alt Pirineu*) project began in 1985 in the districts (*subcomarques*) of Urgellet and Baridà around the town of La Seu d'Urgell. The 1,097 km^2 study area includes over 140 other small settlements in fourteen municipalities (*municipis*). It is surrounded by mountain ranges, mostly rising over 2,000 m in height: to the west, the San Joan de l'Erm (highest peak: Pic de l'Orri, 2,440 m) and Salòria (Pic de Salòria, 2,789 m); to the south, part of the pre-Pyrenean Serra del Cadí (Canal Baridana,

Figure 6.8 Part of the valley of Castellbò, with some lower fields still used for grass and forage and abandoned fields and pastures on upper slopes. The main valley is that of the Segre River, with the snow-covered face of the Serra del Cadí (2,647 m) behind (photograph by Xavier Campillo-Besses)

2,647 m) (Figure 6.8) and the Port del Comte (Pedró dels Quatre Batlles, 2,332 m); and to the north, on the border with Andorra, mountains reaching to 2,898 m (Tossa Plana de Lles). The lowest point (600 m) is where the Segre River leaves the area to the south.

In recent decades, the area has been characterized by many of the typical problems of the Spanish mountains: a major change in the economy, with a change from subsistence to market-orientated agriculture linked to the abandonment of pastures, fields and villages; rural depopulation, especially in more remote and higher-altitude settlements (Figure 6.9) – from 1857 to 1986, the population of the rural areas decreased from 19,905 to 4,470, while that of La Seu d'Urgell increased from 3,792 to 10,101 (Figure 6.10); an ageing rural population; and an inadequate physical infrastructure. However, the area has two atypical characteristics. The first is the long existence of an agricultural economy based on the raising of cattle for milk and beef. This began in the early decades of this century, after the phylloxera virus had wiped out a marginal, but long-established, wine industry; and expanded its dominance as wheat cultivation, sheep raising for meat and wool, and timber production became uneconomic. A second, unique, characteristic is the presence of the principality of Andorra to the north.

Figure 6.9 The village of Turbiàs (1,223 m, 11 inhabitants). A typical village with stone houses and slate roofs. Most orchards have been abandoned, as well as upper pastures, which are being invaded by forests because of low grazing pressure (photography by Xavier Campillo-Besses)

Consequently, smuggling has long been important in the economy and, in recent years, many local people have found work in the service and construction sectors of Andorra, which also provides an important market for food. It is estimated that 30 per cent of the area's economy is directly linked to that of Andorra.

The origins of the project can be traced back to the foundation of the *Grups de l'Alt Pirineu* (Group for the High Pyrenees: GAP), a lobbying organization for the preservation and promotion of the mountains, in 1977. GAP was instrumental in the inclusion of articles giving special recognition to the mountains in both the 1978 Spanish constitution and the 1979 Statute for Catalonia. These led, respectively, to the 1982 Spanish law for agriculture in the mountains and a 1983 law for the mountains of Catalonia, which requires that 5-year plans be developed for each mountain region (*comarca*). During the late 1970s and early 1980s, members of GAP and scientists in the Department of Geography at the Autonomous University of Barcelona (AUB) prepared a number of studies on the Catalan Pyrenees (Campillo et al., 1985). An important result of the close relationships that developed between GAP and AUB was the establishment of 1-week summer courses on Pyrenean studies. These courses, supported by the administration

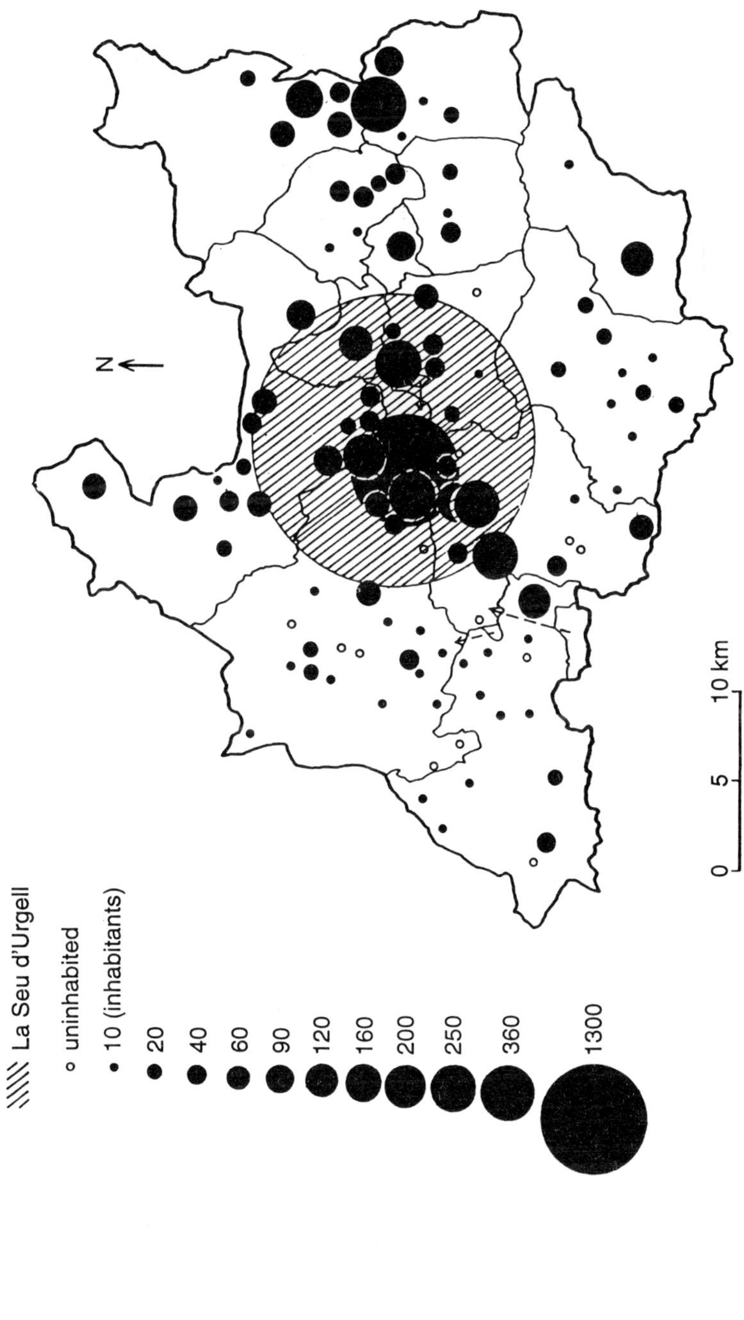

Figure 6.10 Distribution of population in the High Catalan Pyrenees study area in 1981 (from Campillo et al., 1985, with permission)

of La Seu d'Urgell and involving faculty members from AUB, other Catalan universities and, in some years, foreign institutions, have taken place every summer since 1982.

In 1983, a number of members of GAP took part in the international MAB-6 workshop in Château d'Oex, Switzerland, and decided that the approach developed there would be appropriate for the Catalan Pyrenees. As a result, members of GAP and the Department of Geography at AUB submitted a proposal for a MAB-6 project to the Direction for Environment in MOPU later that year. After MOPU approved funds for the project in 1985, the project co-ordinators successfully asked the National MAB Committee to formally endorse it. From 1988, additional funds were also provided by various agencies within the Ministry of Agriculture. It should be noted, however, that the provincial government, not national agencies, sets and funds policies for regional development.

The project had two co-ordinators, one from GAP and one from the Department of Geography at AUB, and received institutional support from the Administration of La Seu d'Urgell, where a secretariat with two full-time positions was established. The co-ordinators were also part of an advisory committee which included representatives of all of the supporting agencies and GAP. However, this committee met rarely. Research was undertaken by members of the Department of Geography at AUB, faculty from the School of Professional Training in Agriculture at Bellestar, and forestry and highway engineers from provincial agencies (some of whom were members of GAP). In total, about twenty-five people were directly involved in the project, as well as many more students from AUB and local people.

The objectives of the project, which ran from 1985 to 1989, were to develop an understanding of the state – and past history – of the region through a series of linked studies, in order to propose solutions for specific problems and develop a regional development plan. To assist with these objectives, the spatial data were compiled in a GIS. It was hoped that the data collected during the study would be used in the creation of a simulation model of the area, based on linear programming, but this never took place because of inadequate financial and human resources.

The results of the project were published as a series of eighteen reports, in both Catalan and Spanish, and in a special issue (no. 12, 1988) of *Documents d'Anàlisi Geogràfica*, published by the Department of Geography at AUB. This is in Catalan, with English, French and Spanish abstracts. In addition, the conclusions of the studies were publicized through a newsletter, distributed to all households in the study area. A short final report, prepared in 1989, was published in Spanish in 1993 (Campillo et al., 1993). The description of the results of the project, presented below, is based on the eighteen reports and interviews (see Author's Preface and Acknowledgements).

The reports can be divided into four groups: background, demography, land use (forestry and agriculture) and infrastructure. The background reports include a general description of the biophysical characteristics, population and economic activities of the area (Campillo et al., 1985) and extracts from historical documents from 1519 to 1925 (Campillo and Sanclimens, 1986). Two reports consider the demography of the area. Campillo et al. (1987) described the evolution of the population and employment structure from 1950 to 1986, with projections to 2000 (Figure 6.11). They also considered commuting and second homes, which have increased rapidly in number and constitute about 40 per cent of the housing stock outside La Seu d'Urgell. Villaró and Campillo (1988) studied rural depopulation (abandonment, seasonal occupation, commuter housing) – its causes (economic, physical infrastructural, social) and consequences (loss of culture and heritage reforestation) – and discussed future perspectives.

Nearly half of the reports are on land use in historical, current and future perspectives. The historical study (Martinez, 1988) considers changes in land use from 1957 to 1987, through the analysis of aerial photographs and maps using a GIS. The analysis showed that the major trend of change was the spread of extensive agriculture (mainly cattle grazing), with some intensification, primarily the irrigation of valley pastures for milking cows. Considerable areas of agricultural land had been abandoned, and often invaded by forests; existing forests had become more dense. The history of the forests was described in more detail by Maza and Aspinach (1987), as part of their overall study of forests, which cover 90 per cent of the study area. This is a detailed study of the functions, structure, ownership and management of the forests; the generally unprofitable market for forest products; and European Community forestry policies; and also includes proposals for planning and management and market improvement. A complementary report represents a methodology for preventing and extinguishing forest fires (Farrero and Villamuera, 1987). A detailed survey of over half the agricultural enterprises was made (Alcala and Nunes, 1988), providing data for the four following studies. Mill and Valls (1988a,b) produced agricultural censuses and economic cost–benefit analyses for the *municipis*. Tulla (1988) studied the structure and transformation of agriculture and developed a typology of enterprises. The data collected during the survey were also used later in a study of the role of women in agriculture (Tulla, 1989). Finally, Sanclimens and Campillo (1986a) described the agricultural policies of the European Communities, but with little discussion of their implications for the study area.

The studies on infrastructure consider transportation, telephones, energy, tourism and municipal services. Gurrera et al. (1987) undertook a multi-criteria analysis of the road network, with particular attention to accessibility, and present proposals for its improvement. Gurrera (1988) described the

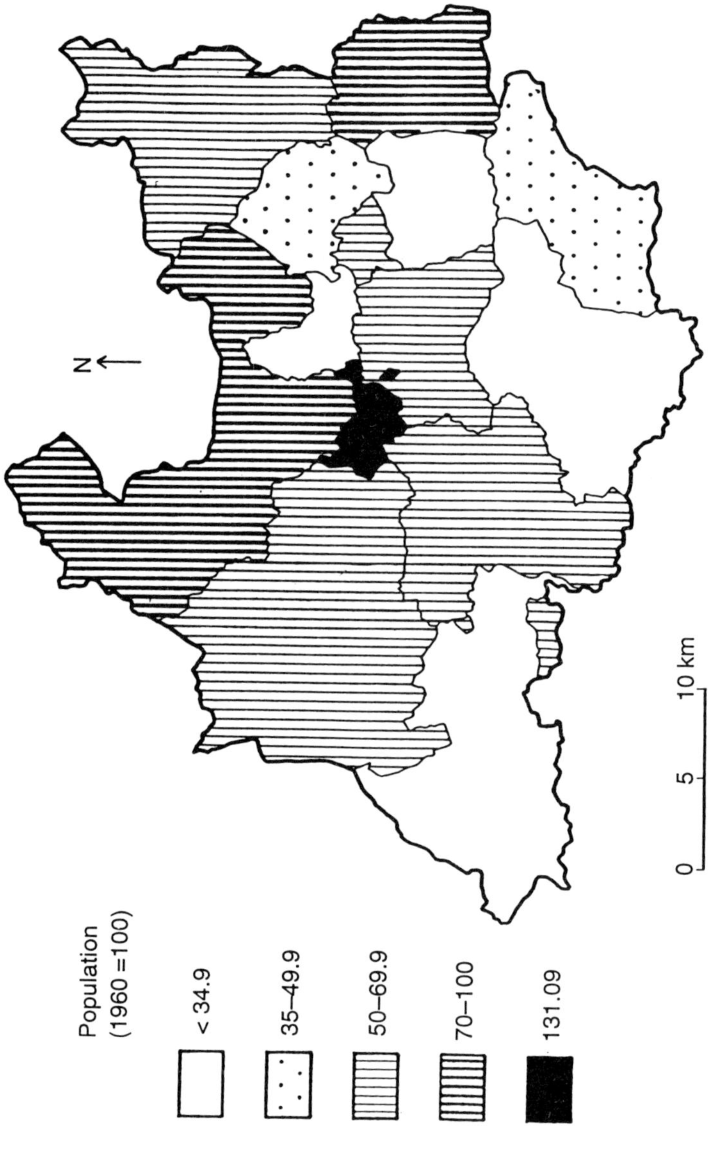

Figure 6.11 Evolution of population in the High Catalan Pyrenees study area in 1981 (from Campillo et al., 1985, with permission)

public transport system in relation to the existing and projected population and infrastructure and proposed future developments based on mobility analyses and estimations of future demands. Similar studies consider the telephone network (Sanclimens and Campillo, 1986b) and energy, with particular attention to renewable energy (Corominas and Puig, 1988). The study on tourism (Sanclimens et al., 1987) focused on the existing facilities and potential for cross-country skiing at the four existing resorts and within the study area in general. It also discussed accommodation and other potential natural and cultural tourist attractions, and presented recommendations for the future of tourism in the area. Finally, the legal and administrative services of the area's small municipalities were assessed by Mateu (1988), who concluded that, as they are often incapable of fulfilling all their legally-mandated responsibilities, new regulations, boundaries and interadministrational co-operation are required.

A brief review of the programme (Ganyet and Tulla, 1988) notes that the dramatic changes in land use and demography are the most visible expressions of the destruction of the socio-economic and cultural structures of the area. They conclude that a specific policy for the Catalan Pyrenees is required in order to integrate these marginal areas – which include about 20 per cent of Catalonia's area, but only one per cent of its population – into a region that is otherwise highly urban and industrialized. However, they note that the implementation of the proposals in the reports, which stress the need for a multi-sectoral economy and an adequately developed infrastructure, require the active participation of local institutions and people.

A number of actions, often implemented by the individuals who undertook the original work in collaboration with local communities and administrations, have taken place since the MAB-6 project, although it cannot be unequivocally said that they resulted from it. First, some of the proposals on the telephone and road networks have been implemented. Second, the Ministry of Culture has funded a 2-year project on the development of multi-sectoral economies in the area and the 5-year comarcal plans have drawn heavily on the results of the MAB-6 project. Third, two existing initiatives whose value was pointed out by the project have been expanded: a co-operative (*mancomunidad*) of the cross-country ski areas and the town of La Seu d'Urgell, to co-ordinate the development of tourism and promote it; and a multi-sectoral economy in the valley of the community of Tuixén, in the south-east of the area, which has stemmed the decline in the population. Finally, new policies for fire prevention and control have been developed and implemented; though these were probably catalysed more by the forest fires of 1987 – which were the most severe for 30 years and burned about one per cent of the forests of Catalonia – than the findings of the MAB-6 project.

There are, however, a few proposed projects which have barely begun or have not yet been undertaken because of lack of funds requested from MOPU, the Ministry of Agriculture, or the National MAB Committee. The former projects include the further development of a GIS to assist in the management of the Cadí-Moixeró Natural Park (41,342 ha). This comprises parts of six *municipis* in the south-east of the MAB-6 study area, and ten outside. The park, established in 1983, has only two inhabited villages within its boundaries, but there are twenty close to it, resulting in considerable pressure for the construction of winter sports areas and second homes. The latter projects include a documentation centre on the Catalan Pyrenees, as well as research on the environmental and social aspects of hunting, recreational uses of water resources (e.g. fishing, canoeing, swimming) and the restoration of the degraded and abandoned lands (involving erosion control, forestry and hydrology).

CONCLUSIONS

The MAB-6 projects in the Aragon and Catalan Pyrenees and Pre-Pyrenees exhibit completely different approaches in spite of broadly comparable trends in economies, populations and resource use in the two study regions. The earlier project, in western High Aragon, was effectively a continuation of the preceding IBP research. It may be described as a broad-based multidisciplinary research programme, which included almost all of the work conducted by the scientists of the CPBE/IPE during the 1970s and early 1980s. The end of the project is hard to identify; research continues on many of its themes. Relatively little of the research was applied, and the project appears to have had few direct influences on policy development either in the study region or at wider scales. While the project led to the compilation of a remarkable volume of data from diverse disciplines, its potential for developing a greater understanding of the interactions of the different human and environmental systems has barely been realized, as almost no syntheses have been prepared or models developed.

One important reason for the lack of interdisciplinary approaches or synthesis in the western High Aragon project is probably that the project was funded only through the usual CSIC channels, so that there was no support for synthesis from a government agency that primarily funds fundamental research. In contrast, the Catalan project was directly funded by two national ministries. This project was distinctive in many other ways, including its strong policy orientation deriving from its promotion by GAP and action-oriented members of AUB, the direct involvement of local people and authorities, its initial development outside the formal national MAB structure, and its relatively short timeframe. Like the earlier project, it was multidisciplinary, but with a strong emphasis on social and applied sciences and almost no natural science research; a mixture that was almost exactly the opposite of the earlier project.

When considering differences between the two projects, it is important to consider that the late 1970s and early 1980s were a period of great societal change after the death of General Franco. Although the co-ordinators of the studies in the Catalan project had scientific and technical training, they were also committed to the cause of sustainable development in mountain communities. They recognized that this required not only Catalan and national policies that acknowledged the special needs of these communities, but also the involvement of local people in the MAB-6 project and consequent actions. While the project did not achieve all of its scientific goals because of inadequate resources, it achieved some policy-oriented goals, perhaps to a greater extent than the Swiss Pays d'Enhaut project on which it was initially modelled. The fact that more recommendations were not implemented derives from the different political orientations of the local administration and the Catalan government during and since the MAB-6 project. Yet, although the project formally finished in 1989, many of its participants and conclusions continue to have a strong influence on the people and institutions of the mountains of Catalonia.

CHAPTER 7

FRANCE

MOUNTAIN REGIONS

With the major exceptions of the Massif Central and Corsica, most of the mountains of France occupy the eastern and southern frontiers with Andorra, Germany, Italy, Spain and Switzerland (Figure 7.1). They may be divided into two types: middle mountains (*moyenne montagne*) and high mountains (*haute montagne*): the Alps, the Pyrenees and the mountains of Corsica. Furthest north, on the German border, are the Vosges (highest peak: Ballon de Guebwiler; 1,423 m) and, to the south of these, the Jura (Crêt de la Neige, 1,718 m) on the Swiss border. Much of the Massif Central is limestone, but its highest peaks are the volcanic Monts d'Auvergne on its western margin, of which the Puy de Sancy (1,885 m) is the highest. A final group of middle mountains are the limestone pre-Alps and outlying massifs such as the Luberon and Mont Ventoux (1,909 m) in the Alpes of Provence.

The massifs of the Alps stretch from the southern side of the Lake of Geneva along the Swiss and Italian frontiers to the Mediterranean. The northern Alps include the massifs of Mont Blanc (4,807 m) and the Vanoise (Grande Casse, 3,852 m). The southern Alps comprise the Pelvoux (Barre des Ecrins, 4,101 m), the Queyras (Aiguille de Chambeyron, 3,411 m), and the Alpes Maritimes (Mont Pelat, 3,051 m). The Pyrenees occupy most of the border with Spain and Andorra. The highest summit on the French side is Vignemale (3,298 m); the highest peaks of the chain are on the Spanish side. The summit of the mountainous island of Corsica is Mont Cinto (2,710 m).

As a result of the differences in altitude, while only the uppermost parts of the middle mountains are above timberline, considerable areas of the Alps and Pyrenees are above timberline. One should note, however, that the timberline in nearly all French mountain areas has been depressed by

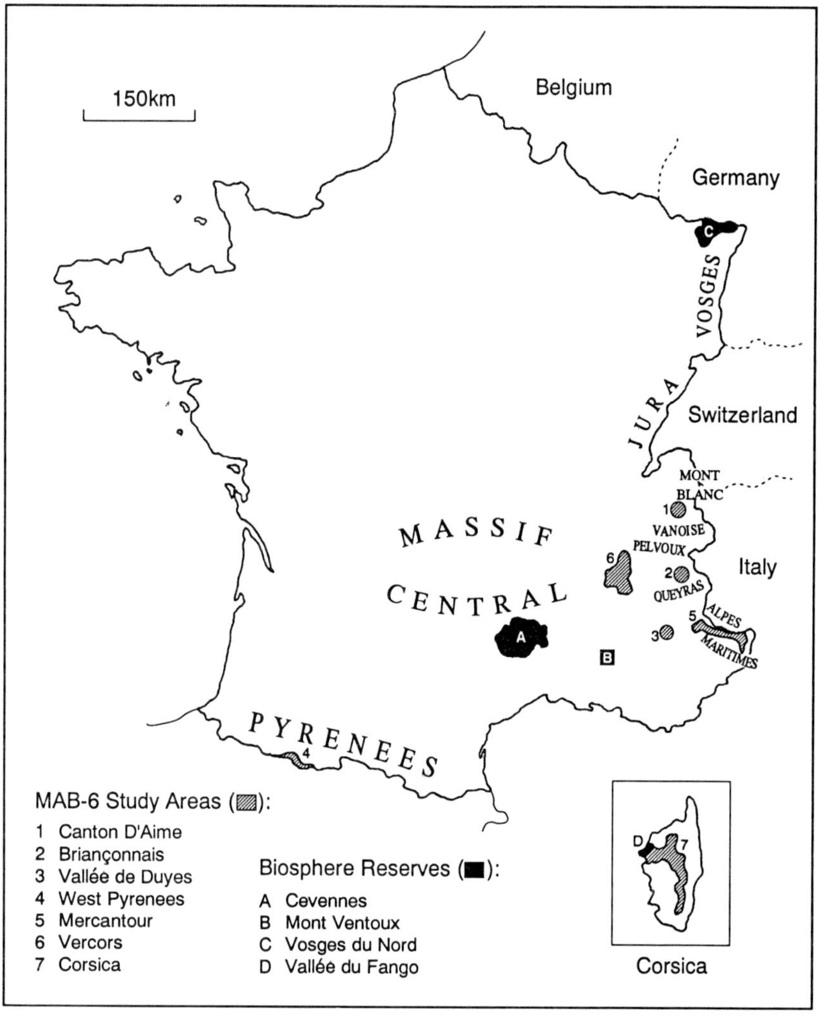

Figure 7.1 France: mountain ranges, MAB-6 study areas and mountain biosphere reserves

anthropogenic action – though considerable reforestation has taken place over the past century. Demographic trends and patterns of economic use vary considerably from one mountain region to another, reflecting very different human histories, economic potentials and biophysical conditions.

The northern and southern Alps have many dissimilarities. The northern massifs have a diversified economy, with a strong influence of tourism (including major tourist resorts used in both summer and winter), a population that is relatively dense and growing, and a well-developed transportation network that provides good access to the major settlements in and near the mountains. However, while this economy is relatively healthy, demands for tourism and the products of some local industries are not increasing to the extent of recent decades and economic development has led to considerable pressures on most environmental resources. In contrast, the southern massifs are characterized by depopulation, summer tourism is more important than winter tourism, and both the economy and transportation networks are weakly-developed. However, environmental pressures are generally less.

The economic and demographic trends characterizing the southern Alps are comparable to those of most of the middle mountain regions and the Pyrenees. In these regions – except in the valleys where tourism has become well-established – depopulation, abandonment of agricultural land (including long-established terraces) and an increasing reliance on sheep and goat-raising are typical. These trends are also found on Corsica, where the problem of increasing fire hazards resulting from the spread of maquis vegetation is common, as in many other Mediterranean mountain areas.

ORGANIZATION OF MAB-6 ACTIVITIES

The French National Committee for MAB was established in 1972 by the Ministry for Foreign Affairs, which appoints the committee's president. The committee includes representatives from a wide range of ministries (and *Directions* within these), research organizations, and national committees and commissions (International Geosphere–Biosphere Programme [IGBP], World Conservation Union [IUCN], SCOPE, UNESCO). The co-ordination of MAB activities is primarily undertaken by the president, working with a small part-time secretariat. This currently includes the president, two vice-presidents, a scientific secretary and the co-ordinators of thematic groups. Since 1982, the secretariat has been established within and supported by the Ministry (sometimes Secretariat of State) of Environment, initially within the *Mission Environnement–Développement* and, since 1991, in the *Direction de la Nature et des Paysages*.

Initially the national committee's thematic groups generally corresponded to the MAB Project Areas. In 1982, five working groups were established. Research in mountain areas was included within the activities of working groups II (disadvantaged, fragile and sensitive zones at all latitudes) and, to a lesser extent, III (conservation of ecosystems, particularly biosphere reserves). In 1986, the working groups were reorganized into six thematic groups, of which 2 and 3 are, respectively, 'rural (non-tropical) terrestrial environments' and 'biosphere reserves'. The stability of the membership of MAB committees and groups from 1982 to 1992 was notable. The committee does not fund research; rather, it encourages interinstitutional interdisciplinary and multidisciplinary research by giving a MAB 'label' to research and other activities. French MAB research projects may be either in France or overseas.

With limited funding from the Ministry of Environment, the National Committee is primarily responsible for promoting research within the framework of MAB and supporting international co-operation. This has included the organization of international meetings, of which two are particularly notable with regard to this report. The first was a colloquium, held in Vallouise in October 1981, on scientific research for the better management of mountain environments, at which papers describing all current MAB-6 programmes in western Europe and the USSR were presented (Ministère de l'Environnement, 1985). The second was the third EuroMAB meeting (Strasbourg, September 1991), which was also supported by UNESCO, the Council of Europe, and Directorate General XII (Environment Programme) of the Commission of the European Communities.

The committee has also supported in principle two initiatives for the co-ordination of research in the European mountains. The first of these was the Network for the European Alps (*Reseau Européen Monde Alpin*), established on the recommendations of a meeting of European ministers of research in 1984. The first meeting of the network, in which institutional and scientific partners from seven countries participated, was in 1986. The network's principal purpose was to stimulate scientific interaction and to provide means for communicating the results of research undertaken by network members, particularly to the European Economic Community. Four themes were emphasized:

(1) The effects of frontiers in the Alps;

(2) Major tourist resorts in the Alps;

(3) Modifications of the alpine environment; and

(4) The new alpine hydrology.

A major activity of the network was the creation of the RESALP bibliographic database, based at the Institute for Alpine Geography in Grenoble, with support from the *Centre national de la Recherche Scientifique* (CNRS). During the late 1980s and early 1990s, many meetings on the four themes were held in different Alpine countries, resulting in the publication of various documents, including three major bibliographies (RESALP, 1986; 1988; 1990). In 1990, the individual responsible for RESALP left for another post with the CNRS. The network has more or less dissolved, and the RESALP database is no longer maintained, though the library and computer hardware and software remain in Grenoble.

The second European initiative supported in principle by the National

Committee was the International Centre for Alpine Environments (ICALPE), established in Le Bourget-du-Lac (Savoie) in 1988, with 3 years' core funding from the *Departement de la Savoie.* ICALPE's primary purposes are to promote the co-ordination of interdisciplinary research in the European mountains and to disseminate the results of such research to wide audiences. An early product was a small book on the Alps, published in English in 1989 and in French and Italian in 1992 (Briand et al., 1989/92). At the second EuroMAB meeting, in 1988, ICALPE took on responsibility for a EuroMAB mountain newsletter; two issues were published, in 1989 and 1990. ICALPE also made an initial survey of the remaining untouched rivers of the Alps for the International Committee for the Protection of the Alps (CIPRA) in 1989, and has published the proceedings of a seminar on Mediterranean mountains and forests (ICALPE, 1992).

ICALPE's main current activity is the organization of three 2-year international interdisciplinary research programmes. Two of these started in 1991. The first, INTEGRALP, concerns the dynamics and management of forests and agro-pastoral systems in the Alps and Pyrenees, and is funded through the European Communities' (EC) STEP programme. It involves fifteen partners from seven countries. The second, FUTURALP, concerns the possible effects of climate change on Alpine ecosystems, and is funded through the EC EPOCH programme (Beniston and Price, 1991; Ozenda and Borel, 1991). It involves seven partners from four countries. In 1991, ICALPE opened an office in Corsica, from which a Mediterranean mountain programme is being organized. The first activity is the co-ordination of MEDIMONT, a 2-year programme on desertification in the Mediterranean mountains, funded by the EC's Environment Programme. MEDIMONT began in 1992, and involves scientists from ten institutions working in five countries. In the absence of reliable long-term core funding, ICALPE's future is unclear. The National Committee is currently reconsidering its relationship to ICALPE.

MAB-6 RESEARCH ACTIVITIES

MAB research activities in French mountain areas began soon after the guidelines of the 1973 Lillehammer meeting were issued, and have now taken place in most of the principal areas (see Figure 7.1). There are four French biosphere reserves which may be described as mountainous. However, research in only one of these is presented in detail below: the Cévennes, established in 1985, with an area of 323,000 ha. Research on Mont Ventoux, designated in 1990 (69,956 ha) has so far been confined to a number of disciplinary studies in the fields of botany, entomology, herpetology, hydrogeology, ornithology and silviculture. Although the

Vosges du Nord biosphere reserve, designated in 1988 (124,000 ha), is quite mountainous, its maximum altitude is 580 m, so that research in the reserve is not included in this report. No research has been conducted in the mountainous part of the Vallée de Fango biosphere reserve, in Corsica, first designated in 1977 and revised in 1990 (23,400 ha). Details of activities in these biosphere reserves may be found in recent reports of the National Committee. These have been published biennially since 1984 (French National Committee, 1984–1992), and also give general descriptions of recent, ongoing and planned research, together with bibliographies of publications deriving from MAB research projects.

Pyrenees

Research in the Pyrenees was proposed by the French delegation at the third session of the MAB-ICC in 1974 (Rey, 1974). This proposal listed a number of ongoing multidisciplinary research programmes and outlined a number of research themes, to be finalized in 1975. Equally, a 6-year integrated pilot project on the effects of human activities on mountain ecosystems in the Pyrenees, starting in 1974, is listed in the 1981 compilation of field projects (UNESCO, 1981). However, no resulting reports could be found. As described below, the West Pyrenees National Park was subsequently one of the research locations of the 1984–1988 interpark programme.

Alps

The first MAB research programme in the French Alps began in 1974 in the Briançonnais, on the Col du Lautaret (Figure 7.2) and the Vallée de la Guisane below it. This region was chosen because it was characterized by most of a range of desired criteria (Guet, 1974):

(1) A representative range of the development of human activities, of which some strongly affect the ecology and/or economy of the region;

(2) The possibility to undertake research expeditiously – based where possible on existing knowledge – which could be used to define operational management strategies for development;

(3) The local or nearby availability of research institutions already undertaking similar research; and

(4) A favourable local institutional climate for undertaking research and acting upon its conclusions.

Figure 7.2 Meteorological station on the Col du Lautaret, Briançonnais

The Briançonnais research programme took place from 1974 to 1980, under the aegis of the *Délégation Générale à la Recherche Scientifique et Technique* (DGRST). Scientists participated from many institutions: CEMAGREF (*Centre national du Machinisme agricole, du génie rural et des Eaux et Fôrets*), CNRS at Grenoble and Paris, INRA (*Institut national de la Recherche Agronomique*), and the universities of Grenoble and Paris VII. The primary synthesis report was published in 1981 (Dollfus, 1981), and a summary can also be found in the proceedings of the 1981 Vallouise symposium (Ministère de l'Environnement, 1985). The 1981 report is divided into five sections: an introduction, including a regional description and an overall summary; the vegetation and climate of the Vallée de la Guisane; aspects of local societal evolution in the Vallée de la Guisane, discussing both pastoralism and tourism; the high-altitude pastoral ecosystem; pastoralism, vegetation and its use for forage and sheep-grazing; an integrated study of the biophysical and pastoral functioning of the Lautaret-Galibier pastures, including geomorphological, climatological and pastoral components.

The second research programme in the Alps began in the canton of Aime (Savoie) (Figure 7.3) in 1981 and concluded in 1984, as an anticipated second phase never took place. The programme was organized by the University of Savoie within the CNRS' *Programme interdisciplinaire de recherche sur l'environnement* (PIREN), and involved thirty scientists, also from the Universities of Grenoble (I and II), Lyon II and Paris XI. The results of the programme, on the management of high mountains and

Figure 7.3 North-west-facing slopes above Landry, Canton Aime, showing partial reforestation of abandoned fields and pastures

environmental consequences, were published in 1985 (Gensac, 1985). This synthesis report begins with a general description of the physical and human characteristics of the area, which includes part of the Vanoise National Park. Subsequent chapters consider: the use of pastures for grazing and the decrease in this use; the forests, their dynamics and management; water resources; slope and soil stability, particularly on abandoned terraces; aspects of the natural heritage; and environmental economics. Figure 7.4 compares the area in 1960 and in 1980, i.e., before and after the development of the local ski area.

Research by scientists from a number of institutions has continued in the area, though not within the MAB framework (Ponge et al., 1993). From 1985 to 1988, the Ministry of Environment supported a project on high-altitude conifer forests, particularly the mechanisms limiting their regeneration.

One emphasis of this research was the influences of bilberries (*Myrtillus* spp.) ericaceous shrubs, and rotten wood on the establishment and survival of spruce seedlings. A second emphasis was the influences of vegetation covers on the evolution of humus at different altitudes, and implications for the sylvogenetic cycle, with further support from the Ministry of Environment from 1991 to 1993.

Inter-park programme (Alps, Pyrenees, Corsica, Vercors)

From 1984 to 1988, a research programme on high-altitude rangelands (*formations paturées d'altitudes*) was undertaken in Mercantour National Park (Maritime Alps), the West Pyrenees National Park, and the Regional Natural Parks of Corsica and Vercors (pre-Alps). The synthesis of results continued through 1990, and resulted in the publication of a summary booklet (Fischesser and Dubost, 1990). Both this booklet and the 1988 and 1990 biennial MAB reports (French National Committee, 1988; 1990) list documents produced during the research phase. The programme was financed by the European Communities, the DGRST, and the Ministry of Environment's SRETIE (*Service de la Recherche, des Etudes et du Traitement de l'Information sur l'Environnement*).

With the general objective of developing better knowledge for the management of high-altitude rangelands, the programme considered two major themes: the evolution of high-altitude rangelands and interactions between wild and domesticated ungulates. Research projects were as follows:

(1) Mercantour: strategies of managing wild and domesticated ungulates to maintain an equilibrium in high-altitude rangelands; effects of declining agriculture and pastoralism on the dynamics of larch forests and on pastoral, animal and landscape resources;

(2) West Pyrenees: comparison of the historical, ergonomic and socio-economic evolution of production systems in the summer pastures of two valleys; spatial and pathological relationships between wild and domestic ungulates; impacts of tourism on chamois;

(3) Corsica: comparison of ecological management strategies for pastures to replace burning; improved management of maquis to provide autumn and winter pasture; scenarios for landscape evolution; and

(4) Vercors: impacts of agro-pastoralism and tourism on the dynamics of wild ungulates; interactions of wild and domesticated sheep; integrated studies for management.

Pre-Alps

MAB research in the pre-Alps has focused on the Vallée des Duyes in the Mediterranean pre-Alps of Digne. The objectives of the first 5-year phase (1984–1988) were:

(1) The development of production systems linking sheep-raising, diversified plant production (*productions végétales diversifiées*) and the improvement of forest resources; and

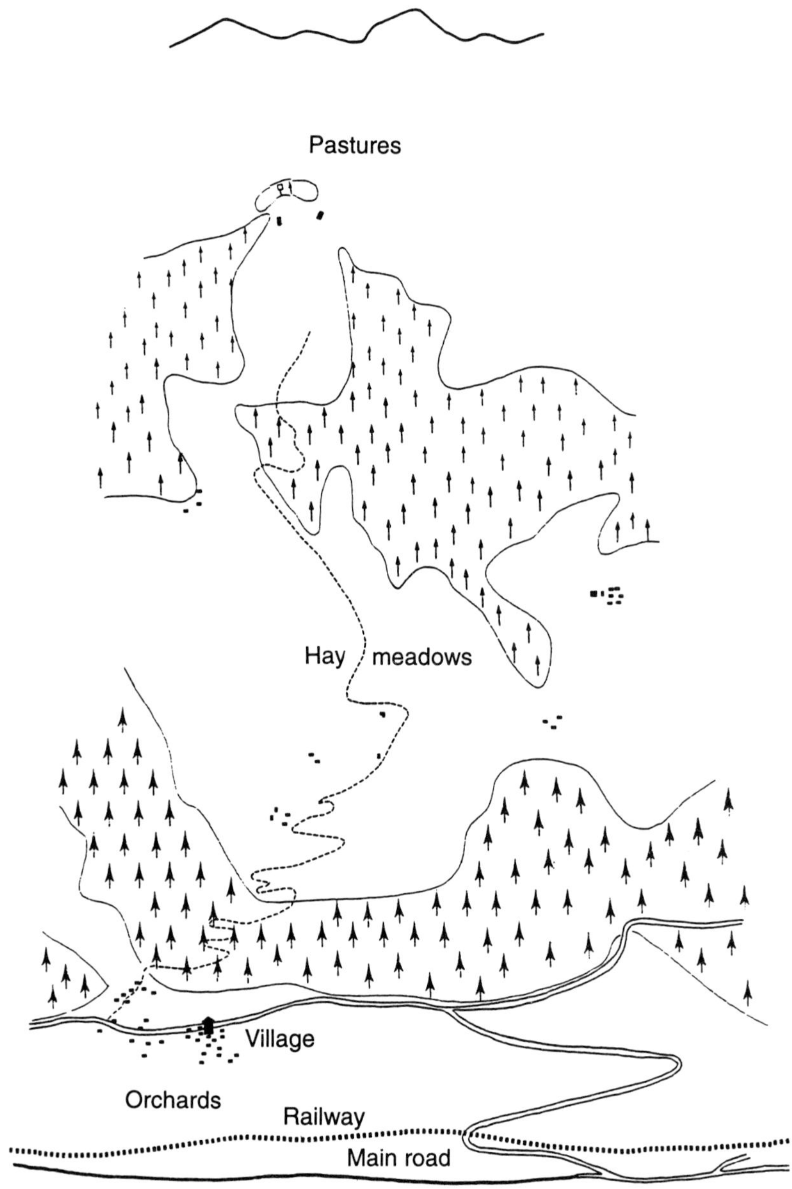

Figure 7.4 Changes in land use on the shaded slope above the village of Vergers, Canton Aime, from 1960 (left) to 1980 (right), associated with the development of La Plagne ski resort (from Gensac, 1985, with permission)

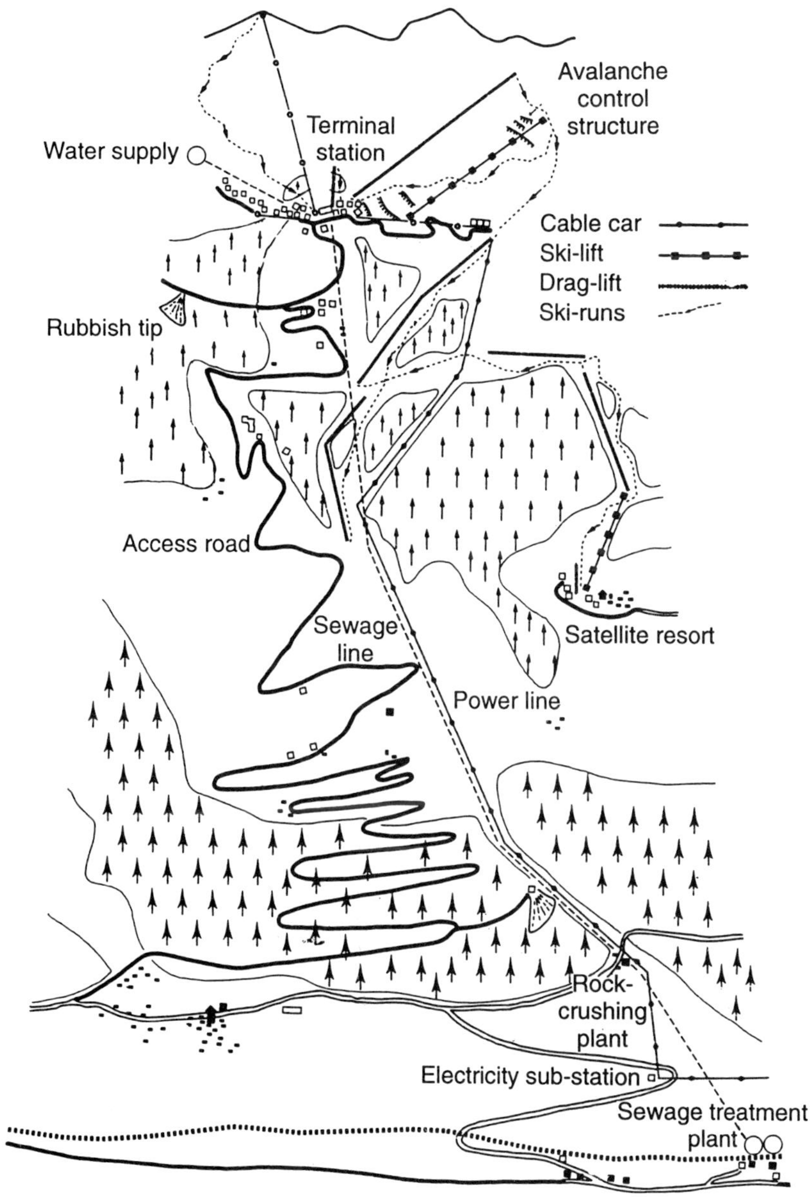

Figure 7.4 *Continued*

(2) Pastoral management ensuring the improvement of forage on grazing land and on oak copses.

In the second 5-year phase (1989–1993), the additional objective of establishing an interactive method for assessing and aiding in negotiations for rural management has been added.

The research has primarily been organized through the Ecodevelopment unit of INRA, at Avignon-Montfavet. Other INRA divisions (rural economics silviculture) have also been involved, together with scientists from CEMAGREF (Aix-en-Provence), the CNRS' *Centre d'Etudes phytosociologiques et écologiques* (CEPE: now *Centre d'Ecologie Fonctionelle et Evolutive* [CEFE]) at Montpellier, the University of Montpellier, and various regional technical and resource management institutions. A considerable amount of the work in the project has resulted, and continues to result, in the preparation of theses.

A first synthesis report was published in 1989 (Msika and Hubert, 1989). After an introduction which presents the history and general methodology of the research, the report includes three major sections. The first considers the farming operations of the area, both in a historical context and with regard to current agronomy, livestock-raising, pasture and land use. The second describes crop and animal production and the management of different types of forage. The third outlines the agricultural system of the region with reference to population, agriculture, the evolution of silvopastoral systems from the nineteenth century to the Second World War and land-use and resource management strategies in the valley.

Massif Central

MAB research in the Massif Central has concentrated on the southeastern Causses and Cévennes massifs. These have similar orographic and mesoclimatic characteristics, but different geologies, human histories and uses of renewable resources. The initial research in this area was a 5-year project, supported by CNRS-PIREN, which began in 1981 and in the Causse Méjan and the Massif de l'Aigoual-Lingas and concluded in 1986. A second phase began in 1987. This has included the preparation of syntheses of the work to date and increasing involvement with management-orientated programmes in the region, particularly in the Cévennes biosphere reserve.

The research has been directed by M. Jollivet (CNRS Sociological Research Group at Nanterre and CEPE/CEFE) and a steering group of five scientists. A large number of institutions – especially from Montpellier, Paris and Toulouse – have been involved. These have included many other CNRS laboratories of the sciences of man and society; agronomic, economic and rural sociological groups from INRA; the Universities of Paris I and VII and Montpellier II and III: the National Agronomic Institute (INA) and the National High School of Agronomy (ENSAM); the Schools of Further Education in Social Sciences (EHESS); the Cévennes National Park; the National Office of Forests; the National Network for Experimentation and Development (RNED); Departmental Directions of Agriculture and Forests; and local agricultural associations and user groups.

The objective of the project has been to establish the scientific foundations for an 'Observatory of ecological, social and economic changes', emphasizing the factors and processes of change which influence the development of natural and altered environments, with respect to societal practices and their appropriate evolution. The 1981–86 project began with a phase of descriptive research, which was followed by experiments and surveys of local decision-makers. It has resulted in many publications and theses, a travelling exhibition entitled 'To be a farmer on a Causse [limestone plateau]: Le Méjan' (Observatoire Causses-Cévennes, 1988), and two summary reports published within the series *Annales du parc national des Cévennes*.

The first summary report considers Le Méjan (Observatoire Causses-Cévennes, 1989). It contains an introduction, eight major papers, a conclusion and a bibliography. The papers consider sheep-raising, land-clearing and associated risks of erosion, the management and quality of pastures and the characteristics of local society. For example, Figure 7.5 shows changes in the membership of the local association from 1970–73 to 1983–86. The second report, on the Aigoual-Lingas massif, has the same structure (Observatoire Causses-Cévennes, 1992). The nine papers consider the heterogeneity of settlement and land-use patterns; changes in patterns of grazing; the physical geography and human uses of the massif; the interactions of agro-pastoralism and erosion; soils, especially for forestry (Figure 7.6); contrasts in forestry management practices; the evolution of agricultural landscapes and their potential for grazing (Figures 7.7 and 7.8); and the protection of nature and agriculture in the National Park. By the end of 1992, a first version of the observatory's baseline database had been achieved. With the intention of assisting ecological management and local development in the region, the database will be used for monitoring forested areas and their use, grazing systems, systems of rural enterprise and rural society.

In 1991, two additional projects supported by the EC's STEP programme and in the framework of the 10th state–region planning contract (1989–93) were given the MAB label. These projects have built on other research conducted in the Cévennes biosphere reserve since 1981, and intensified since 1988 (French National Committee, 1992). The first of the new projects was for 2 years (1991–92), concentrating on three experimental watersheds,

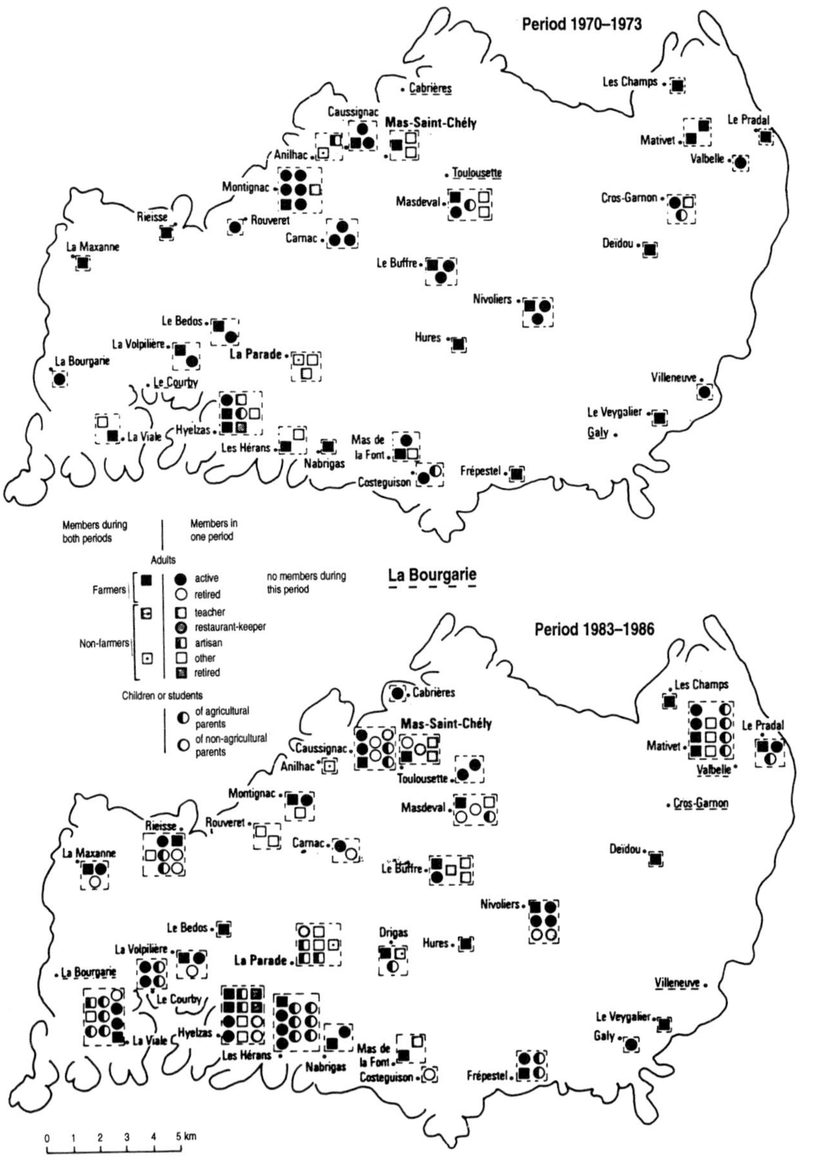

Figure 7.5 Members of the association Le Méjan on the Méjan plateau, 1970–73 (top) and 1983–86 (bottom) (from Observatoire Causses-Cévennes, 1989, with permission)

each with a different vegetation cover, on Mont Lozère in the north of the biosphere reserve. The project, which was part of the EC ENCORE network, was undertaken by CNRS scientists from the Centre for Earth

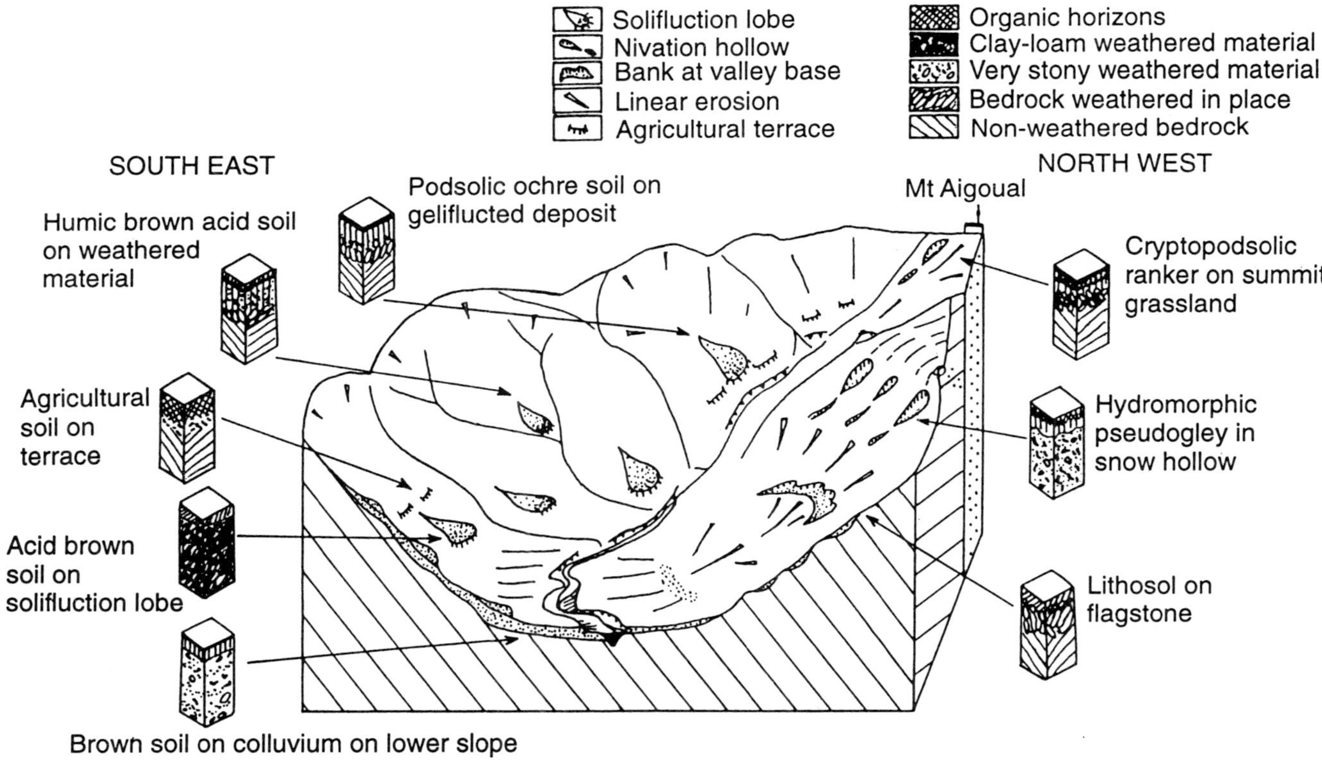

Figure 7.6 Soils and surface features on micaschists on the slopes of Mont Aigoual (from Observatoire Causses-Cévennes, 1992, with permission)

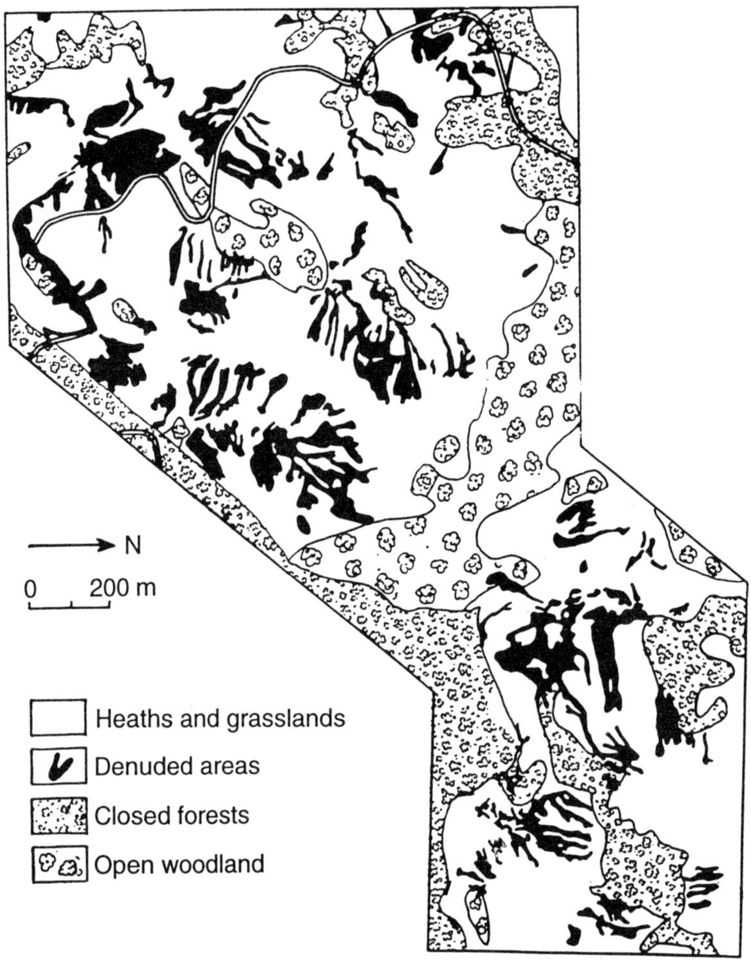

Figure 7.7 Land use on Mont Aigoual, 1948 (from Observatoire Causses-Cévennes, 1992, with permission)

Sciences at Dijon, CEFE (Montpellier), and the University of Orléans. It was an integrated multidisciplinary study, including climate, hydrology, geomorphology, erosion, pedology, hydrochemistry and biogeochemical cycling.

The second project is for 3 years (1991–93). It is a case study of the sustainable development of marginal areas, focusing on chestnut groves, coppices and forests. The project is being undertaken by scientists from CEFE, the Cévennes National Park, the Université Paul Valéry at Montpellier and INRA's Station for Economics and Rural Sociology at Montpellier.

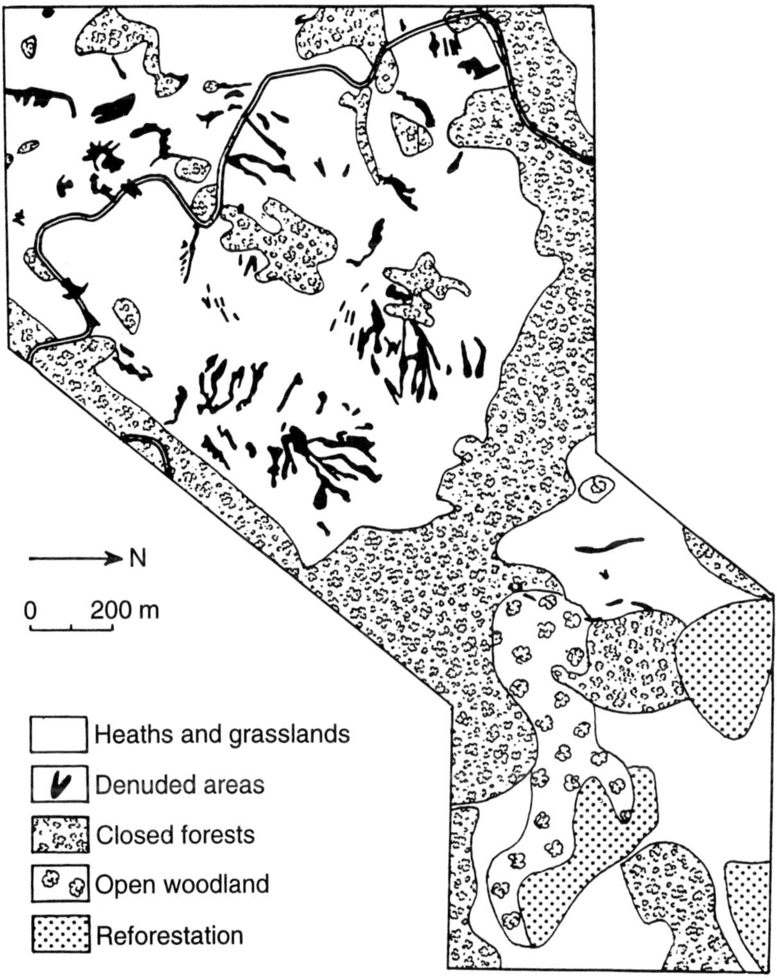

Figure 7.8 Land use on Mont Aigoual, 1981 (from Observatoire Causses-Cévennes, 1992, with permission)

Topics of research include the study of biological diversity in relation to different management regimes, socio-economic studies to identify dominant modes of functioning of agricultural enterprises, assessment of the transferability of results, the study of alternative forest management regimes and the impact of large game animals on forest regeneration. A number of publications and theses have resulted from both of these projects.

CONCLUSIONS

Almost since the beginning of the MAB programme, projects have been ongoing in the mountains of France. The range of research topics and sites has increased considerably since the reorganization of the National MAB Committee in 1982 with, in the mid-1980s, projects taking place simultaneously in as many as eight areas representing the great diversity of France's mountains. In the late 1980s, the number of sites decreased, but increasing emphasis was placed on support for international initiatives which, while they have had many valuable results, have an uncertain future.

It is not easy to judge how many of the research projects and other activities labelled as 'MAB' by the National Committee would have taken place in its absence. However, it seems reasonable to conclude that, particularly since 1982, the programme has had a catalytic effect in providing an extra degree of support to research projects initiated primarily by other national government bodies – such as CEMAGREF, CNRS (especially PIREN and CEPE/CEFE), DGRST, INRA, SRETIE – as well as many universities and the European Communities. The range of institutions involved in these projects has been remarkable, and perhaps this has been one of the greatest successes of the MAB programme in the mountains of France. Yet, in spite of the wide range of disciplines involved and topics considered in these projects, few of them appear to have been especially interdisciplinary, and there has been a strong emphasis on descriptive research.

As models of the interactions of resident and visiting human populations with mountain environments have rarely been developed in the course of MAB-6 projects, their results tend to be specific to the individual study regions. In general, the final reports tend to comprise papers collected together with little attempt at synthesis or application for resource management, although this was often a stated goal. There are some exceptions, notably the comparative interpark programme, which has provided some transferable findings, and recent components of the projects in the Vallée des Duyes and the Causses-Cévennes. It is to be hoped that these recent trends will be continued into the future of the MAB programme in France, particularly through the integration of research conducted by scientists from diverse institutions with the development of monitoring and management strategies by the personnel of the many protected areas in France's mountains.

PART II

MAB in the Mountains of Eastern and Central Europe and the former USSR

CHAPTER 8

EASTERN AND CENTRAL EUROPE AND THE FORMER USSR: AN INTRODUCTION

In contrast to the strong representation of national experts from the mountain countries of western Europe at the MAB-6 Salzburg and Lillehammer meetings in 1973, no experts from eastern or central Europe were present. Thus, it appears that scientists working in mountain areas in these countries had not been incorporated into MAB National Committees at this date. However, a representative from the USSR Academy of Sciences attended both meetings. In his statement at Lillehammer, he noted that investigations on questions relevant to MAB-6 were being undertaken by a number of organizations in various parts of the USSR, and that others were planned (UNESCO, 1974: 89–90).

During the second half of the 1970s, MAB-6 projects began in eastern and central Europe. By late 1975, the Polish National Committee had prepared an interdisciplinary group for studies in the Carpathian mountains, and was seeking to develop a co-ordinated regional research programme with Czechoslovak, Romanian and Soviet scientists (UNESCO, 1976). In 1977, the first meeting of MAB National Committees of socialist countries (except Romania) took place in Moscow (SocMAB). At this meeting, responsibility for the co-ordination of each MAB Project Area was given to a specific country; for MAB-6, to Poland (Gvishiani and Sokolov, 1979).

In the late 1970s and early 1980s, Prof. Zabierowski of Krakow, who was responsible for MAB-6 in Poland, retained responsibility for co-ordinating MAB-6 projects throughout eastern Europe and the USSR (UNESCO, 1978; 1980). Opportunities for such co-ordination were also provided by successive conferences of the MAB Committees of the socialist countries in Warsaw, Poland (1979); Budapest, Hungary (1981); Klinke, East Germany (1983); Sofia, Bulgaria (1984); Brno, Czechoslovakia (1986); Havana, Cuba (1988); and Kiev, Ukraine (1990). Proceedings were only published for the first two of these meetings. At the 1979 meeting, hydrology (particularly the effects of reservoirs), sustainable agriculture and tourism and recreation were identified as priority areas for research in mountain areas (Gvishiani and Sokolov, 1981). Yet, while the need for collaborative studies was emphasized at many of these meetings, substantive co-operative and collaborative activities generally remained limited throughout the Communist era and attempts at co-ordination were given up by the Polish co-ordination centre at Krakow in 1985, when the Institute of Geography of the USSR Academy of Sciences – and specifically the Institute's Laboratory for Mountain Geosystems, formed in 1986 – assumed this responsibility (Soviet National Committee, 1989).

Unquestionably, an important hindrance to the establishment and implementation of multidisciplinary, let alone multinational, programmes was that scientific research, like other sectors of Communist economies, was organized according to the 5-year plans of individual research institutes. These did not allow much of the flexibility required for such complex and dynamic programmes. Nevertheless, work continued within national programmes throughout eastern and central Europe and the USSR, although funds allocated specifically for MAB-6 – or any MAB activities – were extremely limited. The only partial review of this work is provided by the three volumes of proceedings (in Bulgarian, with English summaries) of a major symposium in Vratsa, Bulgaria in October 1983 (Bulgarian National Committee, 1983). However, 134 of the 167 papers presented at this meeting were given by Bulgarian scientists. The majority of the remainder (twenty) were given by Soviet scientists, with nine from four Polish scientists – all describing research in Bulgaria – three from a Romanian scientist and one from a Czechoslovak scientist.

As noted in Part I, since 1987, the biennial EuroMAB meetings have provided further possibilities for interaction between scientists involved in MAB activities in the mountains of Europe. Meetings on the subject of biosphere reserves, beginning with the first international biosphere reserve congress in Minsk, Byelorussia, in 1983 (UNESCO-UNEP, 1984) and followed by both global and regional meetings (starting in 1986 in Ceske Budejovice, Czechoslovakia and most recently in Spindleruv Mlyn, Czech Republic), have also provided opportunities for co-operation in MAB research in central and eastern Europe and the former USSR and scientists from most of the countries in the region have taken part in all of these. Many early biosphere reserves were established in the mountains of this region and have provided a focus for scientific research in these areas.

The need to consider biosphere reserves in this review illustrates the difficulty of undertaking a comprehensive and meaningful review of any of the fourteen MAB Projects in eastern and central Europe. Not only has most of the work in mountains been conducted in biosphere reserves (Project 8), but much of the area of mountain ranges is covered by forests, which fall within Project 2, which considers temperate forests. Much work has been undertaken within this Project in this region, often in mountain areas and, in 1989, representatives of the Czechoslovak and Ukrainian MAB Committees agreed to undertake collaborative work on the impact of human activities on forest ecosystems of the Carpathians. However, little collaborative work has been done, primarily as a result of recent political changes and problems of obtaining financial resources.

For the pragmatic reasons described above, the following chapters are not limited to research that has been labelled as 'MAB Project 6'. They take a broad view of work undertaken in these mountains which has been formally or informally under the auspices of MAB-6 and/or in biosphere reserves in mountain areas.

CHAPTER 9

FORMER SOVIET UNION

Since MAB activities in the states of the former Soviet Union (USSR) took place within the national programme for most of the period under review, this section considers activities within mountain ranges across the former USSR. As discussed below, some co-ordination of activities continues within a number of the states of the Commonwealth of Independent States (CIS), and a conference in October 1992 proposed mechanisms for future co-ordination.

MOUNTAIN REGIONS

In total, the mountains of the former Soviet Union cover a vast area, estimated at 26 per cent of the area of the USSR (12 per cent above 1,500 m) with over 20 per cent of the population (Badenkov, 1992; Figure 9.1). These mountains can be divided into five main areas. The first two are in Europe. Furthest west are the Eastern Carpathians of the southwest Ukraine (highest peak: Goverla, 2,061 m), meeting the parts of the chain in Poland, Slovakia and Romania. To the south and east are the Caucasus, including the highest mountain in Europe, Elbrus (5,642 m). The northern side of this chain is in the Russian Federation, including parts of the Krasnodarskiy Region (*Kray*) and the republics of Kabardino-Balkaraya, North Ossetia, Checheno-Ingushetia and Daghestan. In addition to Elbrus and Kazbek (5,033 m) (Figure 9.2), there are many peaks over 4,000 m along the border between the Russian Federation and Georgia and Azerbaijan. The westernmost extension of the Caucasus is the Ukrainian Crimea (highest peak: Roman-Kosh, 1,545 m). To the south of the main Caucasus are the Trans-Caucasian mountains of Georgia, Azerbaijan and Armenia (highest peak: Aragats, 4,090 m), which extend to the international border with Turkey and Iran. The Urals, with a maximum elevation of 1,895 m

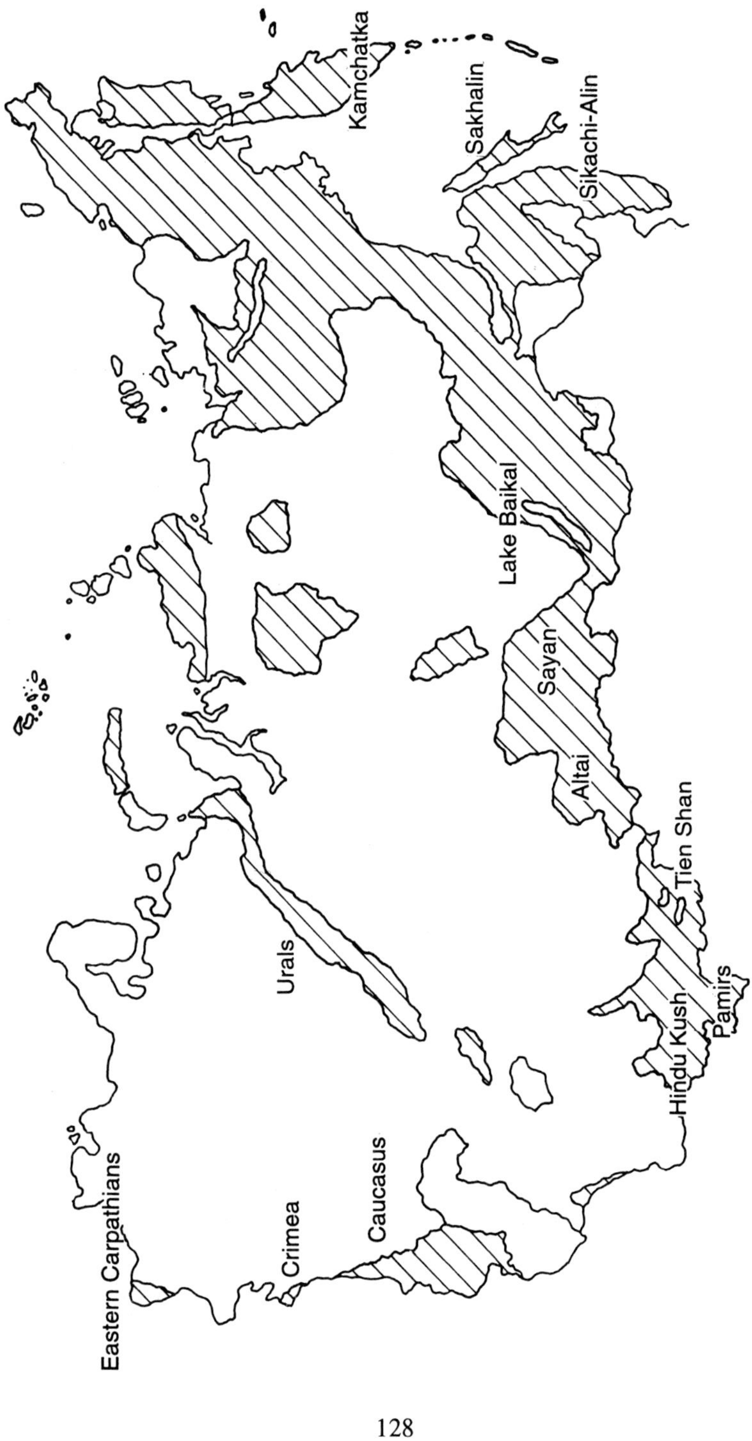

Figure 9.1 Former Soviet Union: mountain ranges

Figure 9.2 Kazbek (5,033 m) and the village of Kazbeghi, Georgia

(Narodnaya), form the divide between the European and Asian (Siberian) parts of the Russian Federation.

Mountains form most of the southern and eastern borders of the former Soviet Union, and cover almost all of the Central Asian republics of Tajikistan and Kyrgyzstan and all the eastern parts of Uzbekistan and Kazakhstan. The Pamir-Alai of Tajikistan (highest peak: Peak Communism, 7,469 m) form the border with Afghanistan, Pakistan, India and China (Figure 9.3). To their north, the Tien Shan (Pobedyi, 7,437 m, in Kyrgyzstan) cross the border from Kyrgyzstan and Kazakhstan into China. There are also a number of lower mountain ranges in Kazakhstan, Turkmenistan and Uzbekistan. At the junction of Kazakhstan, China, Mongolia and the Russian Federation are the Altai mountains, which include Siberia's highest peak (Belukha, 4,506 m) (Figure 9.4).

The Altai provide a transition between the Central Asian mountains and those which cover much of the rest of east and south Siberia, both along the Mongolian and Chinese borders and in the interior, stretching to both the Arctic and the Pacific Oceans. Among these mountain ranges, little known except to scientists and climbers from the former USSR, are the Sayan and Tuva mountains (highest peak: Mongun-Taiga, 3,970 m); the mountains around Lake Baikal (Kodara, 2,999 m); the North-east Siberian mountains (Pobeda, 3,147 m); and the mountains of the Far East, including those adjacent to the Amur and Okhotsk Seas (Taikanskiy Ridge, 2,384 m; Yudomskiy Ridge, 2,889 m), the Sikachi-Alin (Tardoki-Yani, 2,077 m),

Figure 9.3 Fedchenko Glacier, Academy Range, Pamirs, Tajikistan (photograph by Jack D. Ives)

Sakhalin (Lopatin, 1,609 m), and Kamchatka (Kliuchevskaya Sopka, 4,750 m).

In summary, the mountains of the former Soviet Union are at latitudes from arctic to semi-tropical, with climates from maritime to extreme continental, and thus exhibit a marked diversity of physical environments. Similarly, their histories of human use vary greatly. These have been particularly long in the Carpathians, the Caucasus and Central Asia, all of which have been characterized by traditional patterns of agriculture in valleys and seasonal grazing in subalpine and alpine zones. However, overgrazing and monoculture agriculture in the mountains of the Caucasus and Central Asia in recent decades have led to major decreases in productivity. Deforestation, leading to soil erosion, desertification and other environmental changes including catastrophic landslides, has been a major problem in many regions, particularly arid areas, e.g. Uzbekistan and Kyrgyzstan. High rates of population growth, particularly in Islamic

Figure 9.4 Belukha (4,506 m) (photograph by Ian Heywood)

republics, also lead to increasing pressures on the environment. Conversely, the emigrations, often forced, of mountain people which took place from the 1930s to the 1970s from the northern Caucasus, Georgian, Azerbaijani and Central Asian mountains have only recently begun to be reversed. In all of these mountain regions, many rivers have been dammed for irrigation and hydroelectricity, and more dams are under construction, mainly in Central Asia, and planned, particularly in the Caucasus. Tourism is an increasingly important source of environmental and socio-economic change in the Caucasus and Kyrgyzstan, and also in many other regions.

While the Altai and some of the other southern Siberian mountains have a long history of grazing by the animals of both settled and nomadic tribes, human settlement and use in most of the Siberian mountains have been greatly limited by their severe environments and, particularly, their harsh climates. Only in this century have the mountains of north-east and Far East Siberia begun to be developed for mining; forestry has also become a major industry in the Far East. While tourism is also growing in the Siberian mountains, numbers of tourists are generally very low, especially because of problems of accessibility. However, tourism is widely seen as a catalyst to economic development, and many projects are underway or planned, especially in the Far East (Kamchatka, Sakhalin) and the Altai. In addition, while relatively few dams are currently operational in mountain areas, many are under construction or planned in most Siberian mountain ranges, except Kamchatka where geothermal power is available.

ORGANIZATION OF MAB-6 ACTIVITIES

The Soviet National Committee for MAB was formed in 1975 by a decision of the USSR State Committee for Science and Technology, the USSR Ministry for Foreign Affairs and the Presidium of the USSR Academy of Sciences, as an interdepartmental body of the USSR Commission for UNESCO and the USSR Academy of Sciences. The A.N. Severtsov Institute for Animal Evolutionary Morphology and Ecology of the USSR (now Russian) Academy of Sciences has acted as the Secretariat for the National Committee since 1975 (Soviet National Committee, 1989). Initially, responsibility for co-ordinating Project 6 activities across the USSR was given to the Institute of Geography of the Georgian Academy of Sciences, in Tbilisi. In addition to these all-Union structures, MAB Committees, each with a sub-committee for MAB-6, were established in each republic. Each year, a meeting of these committees was held in a different republic.

As described in the following section, MAB-6 activities in the late 1970s and early 1980s were limited, being conducted by small, isolated groups of scientists. While strong possibilities for co-ordination existed until 1979, these effectively ceased with the death of the Director of the Georgian Institute of Geography. His successor was, unfortunately, less active and did not have such a wide network of contacts. Thus, in 1983, the Soviet National Committee asked the Institute of Geography of the USSR Academy of Sciences, in Moscow, to take over responsibility for co-ordinating MAB-6 activities. This institute already had a long experience of work in mountain areas and had strong connections to scientists working in European mountains.

In 1985, working groups for the co-ordination of MAB-6 activities were established for six regions: Altai-Sayan, Carpathians-Crimea, Caucasus, Central Asia, Siberia and the Far East and Urals. These groups were to co-ordinate work on the following themes (Badenkov, 1987):

(1) Theoretical bases of the development of mountain regions;

(2) Patterns of the formation and development of natural complexes and their components in different altitudinal zones;

(3) Scientific substantiation of rational use of natural resources and nature conservation;

(4) Natural hazards;

(5) Problems caused by the development of agriculture, industry, transport and different land-use systems;

(6) Environmental monitoring;

(7) Recreational use;

(8) Impacts of economic activities and problems of ecological–economic planning; and

(9) Systems of settlement and human ecology.

The MAB-6 network encompassing the regional working groups was strengthened by annual meetings and the publication of an annual bulletin of activities, beginning in 1985. In 1986, a new Laboratory of Mountain Geosystems (since 1992, International Mountain Research and Information Laboratory: IMRIL) was established within the Soviet Academy's Institute of Geography, and took over the functions of a MAB-6 Secretariat. Nevertheless, one should note that funds for these functions, provided by the Academy of Sciences (originally USSR, now Russian) – and more recently also by the Russian Ministry of Ecology – have been very limited.

The structure established in 1985 continued with some success until the disintegration of the Soviet Union in 1990, when the last annual meeting was held, and the last bulletin published. AT present, IMRIL retains the functions of the Secretariat for MAB-6 in the Russian Federation, and maintains links to scientists undertaking MAB-6 activities in other states of the CIS. However, official links to the Ukraine and the Caucasus are limited by strong nationalist governments and political instability. In October 1992, a conference, supported by the MAB-6 committee of the Russian Federation, was held in Vladikavkaz, North Ossetia (Russian Federation). About 150 scientists from across the CIS attended, to present scientific papers and review the current situation. The participants in the conference adopted a resolution establishing a Eurasian Mountain Association, with the objective of continuing and expanding collaboration between scientists working in the mountains of the CIS. However, the Association exists only in principle; the resolution is unlikely to be implemented without an improvement in the present unstable political and economic situation.

MAB-6 RESEARCH ACTIVITIES

At the 1973 Lillehammer meeting of the international Working Group on MAB-6, the representative of the USSR Academy of Sciences noted that many organizations were conducting scientific investigations relevant to MAB-6 in the Russian Federation and the Armenian, Georgian and Uzbek republics. Multi- or trans-disciplinary investigations of the natural resources of mountain ecosystems were planned, paying attention to the problems of land use, protection of animals and plants and human ecology, and taking into account the effects of technology and tourism. Regions proposed for such studies included Trans-Caucasia, the Komi Republic, the Pamirs, the Trans-Baikal area and the mountains of the North-east USSR (UNESCO 1974: 89–90).

Important published sources for information about projects within the scope of MAB-6 in the former USSR are the annual MAB-6 bulletins published from 1985 to 1990; a listing of scientific researchers and publications, published in 1987 (Sokolov, 1987); a state-of-knowledge review of MAB-6 in the Soviet Union, published in 1988 (Dzhaoshvili et al., 1988); and national MAB newsletters published in 1989, 1990, and 1991. Another early source of information is the 1981 compilation of MAB projects (UNESCO, 1981) This lists eight MAB-6 projects in the Soviet Union beginning between 1975 and 1978, and lasting for periods from 3 to 10 years. While some were site-specific studies on limited themes, most were long-term comparative studies co-ordinated from the Georgian Institute of Geography or the Geography Department of Moscow State University.

These early MAB-6 activities primarily took place in the Caucasus and the mountains of Central Asia. Primary sources for the early MAB work in the Caucasus are books by local scientists (e.g. Kobakhidze, 1985) and deriving from international field seminars in the Caucasus with French and Bulgarian scientists (Gerasimov et al., 1980; Gerasimov and Galabov, 1984). It is notable that the first of these seminars, which took place in 1974, early in the history of the MAB programme, primarily considered natural scientific themes, the one exception being the assessment of anthropogenic changes in vegetation. However, the second, which took place in 1978, gave considerable attention to water resources, industry, agriculture, recreation and population geography. The early MAB work in Central Asia is described in the summaries of papers given at a conference on semi-arid and desert mountains, held in Alma-Ata in 1982 (Chupakhin and Kochurov, 1982), and a number of reports (e.g. Mamytov et al., 1982; Mamytov, 1987). While review of annual reports on the status of MAB-6, and plans for the Project until 1985, suggest that the programme was strong, in reality such reports and plans were merely compilations of projects with 'mountain' themes. Scientists appreciated the 'MAB label' as UNESCO had a high profile in the USSR, and any involvement in UNESCO activities provided a possibility for international contacts.

These early studies (i.e. until the mid-1980s) were generally conducted by small groups of scientists working in relative isolation. While these groups often included scientists from a number of disciplines, each tended to work separately, and thus the general extent of multidisciplinarity was that papers describing research in different disciplines were bound together in reports or books. The establishment of the new structure for co-ordination in 1985 allowed the Director of the Laboratory for Mountain

Geosystems to introduce the integrated concepts of MAB-6, deriving particularly from the Swiss MAB-6 programme and also from contacts with the Institut de Géographie Alpine in Grenoble, to other Soviet scientists (Badenkov, 1990). As a result, research by scientists of different disciplines – primarily human and physical geographers, but also biologists and economists, among others – began to be integrated to some extent although, as before, many 'MAB-6' research activities were only labelled as such; i.e. their conceptual basis was rarely, if at all, interdisciplinary or integrated. The research which began in 1985 with the formation of the regional working groups is described below on a regional basis, based primarily on the unpublished annual reports prepared from 1985 to 1990 and the directory of the USSR MAB scientists published in 1987 (Sokolov, 1987). These documents are the source of the remainder of this and the next section of the report if no other source is given.

In addition to these documents, a number of books and publications deriving from conferences bring together the findings of research by scientists working in mountains across the former Soviet Union. From a conceptual viewpoint, one important product of MAB-6 research is the book *People and mountains (Lyudi i gorni)* (Avakyan, 1989). The books edited by Zaionchovskaya and Polan (1988) and Akhaminov and Polan (1990) are collections of many papers based on MAB-6 research on economic development and settlement in mountains. The majority of papers in both books consider the Caucasus (mainly Georgia and the northern Caucasus), but a few consider the Carpathians, Kyrgyzstan, Tajikistan, and the Altai. The papers in Zaionchovskaya and Polan's (1988) book focus primarily on agriculture and the dynamics and distribution of mountain populations. Akhaminov and Polan's (1990) book also considers these themes, together with a wider range of topics, including forestry, nature conservation, recreation, transport and urban and industrial development.

A number of conferences permitted scientists working in MAB-6 projects across the USSR to present and discuss their research. The first of these, on ecological and socio-economic problems of mountain regions, was in October 1986, in Telavi, Georgia, sponsored by MAB and the United Nations Environment Programme. Papers were presented by Soviet and a few foreign scientists on four general topics: rational land use, recreation, population distribution and industrial development. Among the papers presented by Soviet scientists, about one-third considered the Caucasus, while others considered the Carpathians, the northern Siberian mountains and the Tien Shan, and presented comparisons of the mountain regions of the USSR.

In October 1989, a major international conference on 'The transformation of mountain environments: regional development and sustainability; consequences for global change' took place in Tsahkadzor, Armenia,

Figure 9.5 Participants in the International Conference on the Transformation of Mountain Environments, October 1989, with Elbrus (5,642 m) in the background (photograph by Jack D. Ives)

co-sponsored by UNESCO-MAB, the USSR Academy of Sciences, the United Nations University (UNU), the East–West Center, and the Commission for Mountain Geoecology of the International Geographical Union (Figure 9.5). Many of the papers have been published in volumes 10 and 11 of *Mountain Research and Development*, and UNU were to publish a book based on the conference papers and other MAB-6 research in the former USSR in 1994. The conference was divided into four sections in which papers were given by Soviet and foreign scientists:

(1) Theory, methods, models: evolution and progress;

(2) Implementation of the results of scientific research in practical management: case studies;

(3) Mountain hazards and human activities; and

(4) Regional and global problems: space–time scales and relationships, future research orientations.

In addition, a session was devoted to papers given by Armenian scientists, on problems of mountain regions in their republic. The case studies from the Soviet Union considered all mountain regions, on topics including agriculture, botany, climatology, demography, economics, forestry, natural hazards, nature conservation, pedology, range management, recreation and water resources.

In addition to these conferences with foreign participants and the annual meetings of the national MAB-6 Secretariat and the heads of the regional working groups, two conferences brought together MAB-6 scientists from across the USSR. The first was in Barnaul, Siberia in October 1989, with 120 participants. Papers addressed six main themes: hydroelectric power development and its potential impacts; ecological–economic approaches to the development of mountain regions; assessment of natural hazards; recreational development; industrial development; and climate and glaciology.

A second conference of scientists working in mountains across the former USSR took place in Bishtek, Kyrgyzstan, in October 1991. The abstracts of presentations have been published (Osmonov et al., 1991). Four main themes were considered, each in relation to the overall theme of rational land use in mountain regions: geoecology of mountain landscapes; interdisciplinary studies; economic and ecological aspects of land use; and mapping. The majority of the 138 papers derived from research in Central Asia, particularly Kyrgyzstan, but there were also many papers based on research in all other mountain ranges of the USSR. The papers covered a very wide range of topics, including human impacts on flora and fauna, soil erosion and desertification, recreation and its impacts, natural hazards, agricultural development and interactions with other land uses, water resources, nature protection and environmental monitoring.

Altai-Sayan

MAB-6 activities in the Altai-Sayan have mainly considered climatology, natural hazards and the use of nival–glacial resources, both as water sources and for recreation. A conference held in Barnaul in November 1985, on the role of nival–glacial forms in the dynamics of mountain systems, provided an initial review of ongoing and planned work. While the conference was attended by scientists from across Russia and Kazakhstan, its emphasis was on the Altai-Sayan. This meeting provided the basis for further research during the 1980s. In 1989, the Altai Laboratory of Mountain Land Use was founded in Barnaul, within the Institute of Water and Ecological Problems of the Siberian Branch of the USSR (now Russian) Academy of Sciences. The primary foci of research are the analysis of natural hazards and environmental stability in general and the study of climatological influences on glaciers. Field-work has mainly concentrated on the Ob and Katun river basins, especially in relation to the Katun dam, now under construction.

Carpathians – Crimea

Research within the scope of MAB-6 in the Carpathians and Crimea has primarily been undertaken by Ukrainian botanists, forest scientists, zoologists and pedologists. In addition, some research on nature protection and recreation has been done. The primary emphases of these studies have been:

(1) Forest ecosystems, with emphasis on studies of structure, function, productivity and the impacts of human activities;

(2) Water resources, with particular consideration of the protective role of forests and the effects of different harvesting regimes;

(3) Nature protection; and

(4) In the Crimea, the chemical composition of rainfall, and recreation potential and impacts in forests.

While MAB-6 research in the Carpathians has primarily been within the fields of natural science, many of the findings of this research have been integrated with studies by social scientists. The primary result of this integration, undertaken within the interdisciplinary spirit of MAB, is a series of four books edited by members of the Ukrainian Academy of Sciences. The book edited by Golubets et al. (1988) considers physical geography, forests and their management, agriculture, wildlife, and nature protection. Dolishniy (1988) presents a variety of economic topics, including industrial structure and inter-sectoral linkages, demography and employment structure, regional economic development, interactions between economic development and environmental protection and changes in the quality of life. Goshko et al. (1989) consider ethnography, demography and settlement, industry, transport and other infrastructure, family life, psychology and art. The history of the region is presented by Slivka et al. (1989).

An important recent initiative in the Eastern Carpathians has been the proposal of an international biosphere reserve in Slovakia, Poland and the Ukraine (Stuzica reserve). Although the MAB-ICC has approved only the Polish-Slovak portion of this proposed reserve for the time being, it is expected that the Stuzica area will be a focus for future MAB-6 activities in the Ukraine. MAB-6 activities in the Carpathians and Crimea are now being co-ordinated by the Ukrainian National Committee for MAB. This has published a listing of research themes until 1995 as well as a booklet, prepared for the last meeting of MAB committees of the socialist countries in 1990 (Ukrainian National Committee, 1989; 1990). These documents show that continued research in the fields of study described in the first paragraph was intended. However, financial constraints have limited both this research and the development of a monitoring system for the biosphere reserves of Ukraine, many of which are in the Carpathian and Crimean mountains (Malysheva and Potapenko, 1993).

Caucasus

As described above, the Caucasus has been a major focus of MAB-6 activities since their beginning in the former Soviet Union. In addition to research on a wide range of topics in both natural and social sciences, a number of conferences have been held. When the new co-ordinated structure for MAB-6 was established in 1985, four major areas of scientific research were identified:

(1) The formation and function of mountain geosystems;

(2) Regional land use and the protection of natural resources;

(3) Natural hazards and their prevention; and

(4) Socio-economic processes.

From 1987, this research, by scientists from the Caucasian republics and the Laboratory for Mountain Geosystems (Moscow), was organized as a special project on 'The transformation of environment, economy, and population in the Caucasus.' A preliminary report was published in the proceedings of the major 'Earth Transformed' conference, held in the USA in 1987 (Badenkov et al., 1992; Figures 9.6 and 9.7), and a final report is almost complete.

Most of the regional MAB-6 meetings in the Caucasus have considered work conducted by social scientists. The first conference, on the socio-economic development of mountain regions, took place in Yerevan, Armenia, in 1985 (Vartanian et al., 1985). Its primary focus was on the development of agriculture, particularly from an economic perspective. Most papers concerned Armenia, Azerbaijan and Georgia, while a few described work in central Asia and the Ukraine. A second conference in Yerevan, in April 1987, discussed problems associated with mapping the distribution of population in mountain areas. Most participants were from the Caucasus, a few from Tajikistan. A third meeting, in Dombae, considered

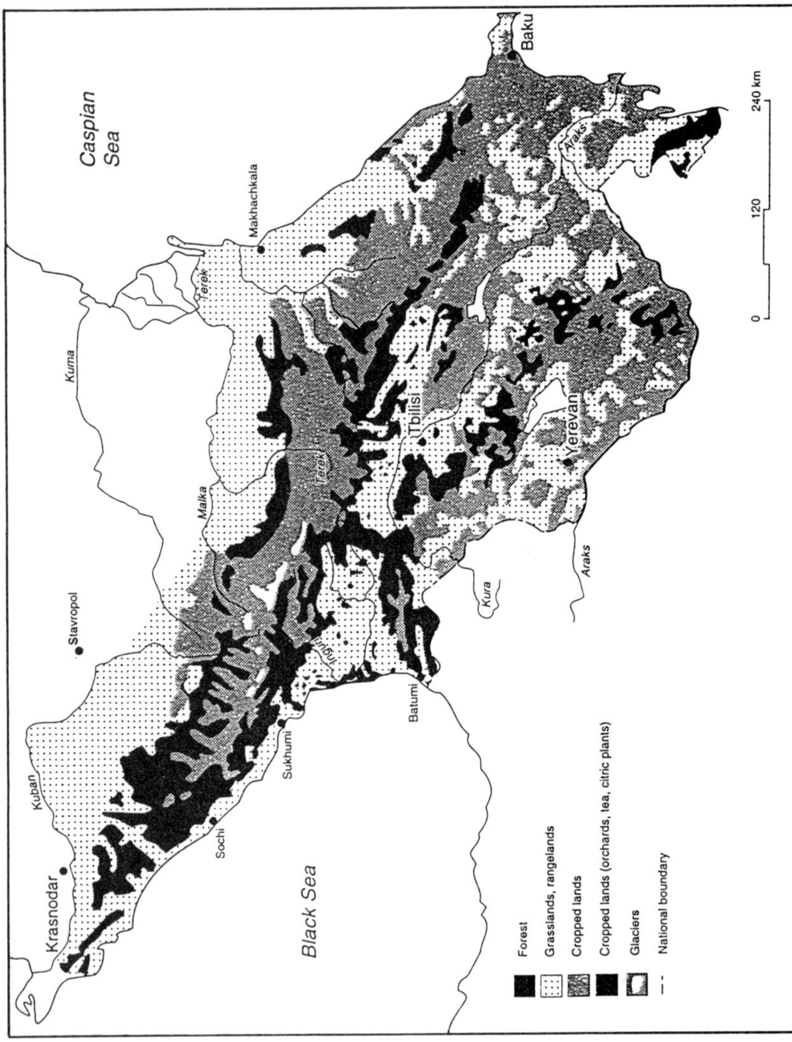

Figure 9.6 Land use in the Caucasus in 1700 (from Badenkov et al., 1992, with permission)

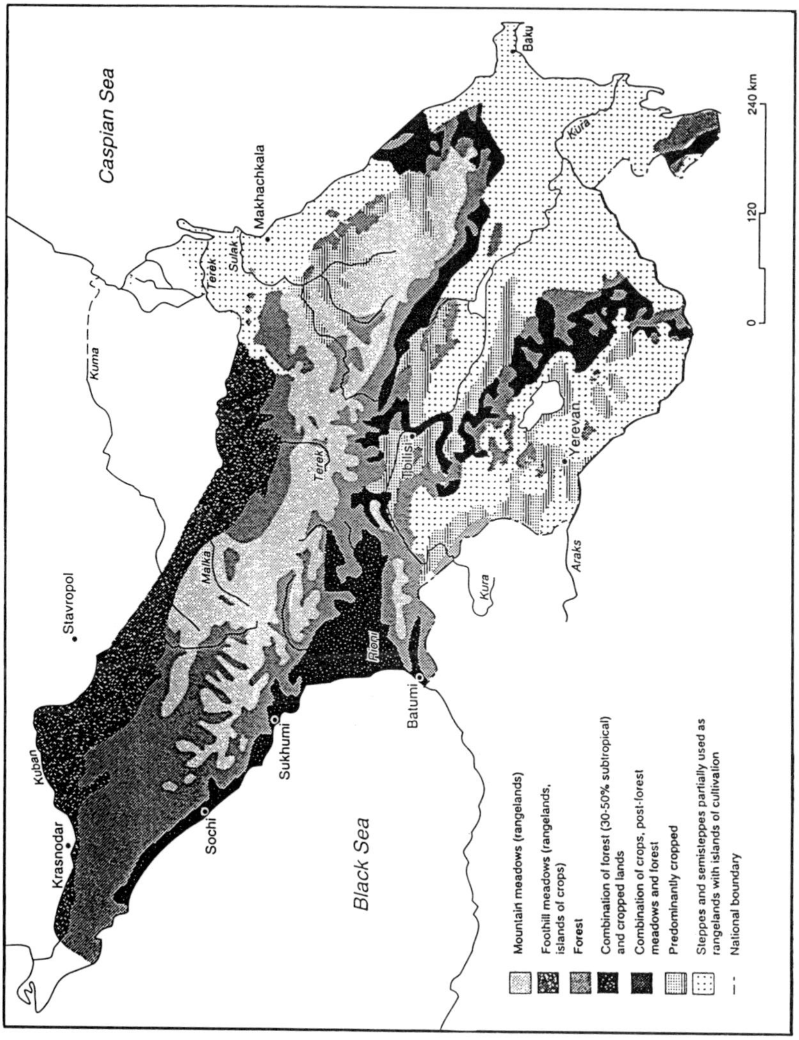

Figure 9.7 Land use in the Caucasus in 1986 (from Badenkov et al., 1992, with permission)

the demography of the northern Caucasus. As a result of this meeting, monographs on demographic dynamics and distribution in each region of the north Caucasus are being prepared.

In addition to the work on agriculture, demography and economics, social scientists working in the Caucasus have considered the development, interactions and impacts of forestry, industry, transport, tourism and urbanization. Papers on many of these topics have been published in a book edited by Kotlyakov and Yashina (1987), which also includes papers on botany, glaciology, hydrology, microclimatology, pedology and zoology. Papers presented at the 1989 international conference in Tsahkadzor provide a later overview of MAB-6 work in the Caucasus.

Most of the work based on the physical sciences has been applied, often including collaboration with social scientists. Topics of research, principally in Georgia, have included:

(1) The water regime and anthropogenic changes in subalpine soils;

(2) The protective function of forests;

(3) Anthropogenic changes in the biology of water and terrestrial organisms, particularly resulting from heavy metal pollution;

(4) The productivity of high mountain meadows; and

(5) Multiple uses of forests (i.e. production, protection and recreation).

Research has also focused on the assessment of natural hazards and their impacts, particularly with respect to central and western Georgia and two specific events: the major rains and snowfalls of winter 1986–87 in the northern Caucasus, and the December 1988 earthquake in Armenia. In the northern Caucasus, water resources have been a major focus of research, resulting in the publication of a book edited by Vladimirov et al. (1991).

The co-ordination of research in the Caucasus has effectively ceased in the unstable political situation deriving from the demise of the Soviet Union. Armed conflict both between and within states and republics, very often in mountain areas, greatly limits possibilities for research. Similarly, an uncertain economic situation means that there are few resources for completing or publishing syntheses of the significant volume of research undertaken in past years. However, the breadth and depth of research conducted within the scope of MAB-6 should provide a valuable baseline for future research and planning when political stability returns to the region.

Central Asia

MAB-6 research has been conducted in all of the former republics of the USSR in Central Asia. Not surprisingly, the two main foci have been Tajikistan and Kyrgyzstan, both of which are primarily mountainous. Two conferences have primarily been based on work in Tajikistan. The first was an Indian–Soviet conference on regional development in mountain areas, held in Dushanbe in September 1985 (Alam and Kidwai, 1987). Papers from Tajikistan considered agriculture, forestry, land reclamation, nature protection, hydroelectric power development and its environmental impacts and water resources. A later conference, in September 1988, specifically considered the integrated development of Badakhstan. This is the most south-easterly portion of the Pamirs of Tajikistan, and is characterized by high population growth rates, limited land for agriculture, and low accessibility, income, education, and energy supplies. However, there is a great potential for natural resource utilization, particularly through the development of hydroelectric power, though this is unlikely in the current era of political instability.

In addition to the development-orientated work described at the meetings in 1985 and 1988, most other research in Tajikistan has concentrated on geobotany, forestry, anthropogenic changes in meadow ecosystems and recreation. In addition, an international project on the 'Nature, Population, and Economy of Tajikistan: An Electronic Atlas-Scenario of Regional Development in the 21st Century' (Tadj-XXI) began in 1991 (Badenkov, 1992; Koshkariov et al., 1994; Figure 9.8). Including research at two scales – the whole of Tajikistan and four test areas – its primary objective is to evaluate whether the environmental and resource capacity of the region will be sufficient to provide living space and meet the demands of a rapidly-growing population. The project is under the auspices of the Life Support Systems Programme of the UNU, which supported field seminars on the region in 1990 (in Switzerland) and in 1992 (in Dushanbe). Scientists from IMRIL, academic institutes and organizations from Tajikistan, and the University of California at Davis are involved. The publication of an English-language monograph is planned at the conclusion of the project in 1994. However, to date, no summary or synthesis of MAB-6 research in Tajikistan has been prepared.

Research in Kyrgyzstan, particularly the Tien Shan mountains, has considered a wide range of both natural and social science topics. The primary published source for this work is the collection of abstracts of papers presented at the conference in Bishkek in October 1991 (Osmonov et al., 1991). A number of studies have described the glaciers, soils, water resources and forest and other vegetation communities of Kyrgyzstan. Many of these studies have provided information used both for assessments

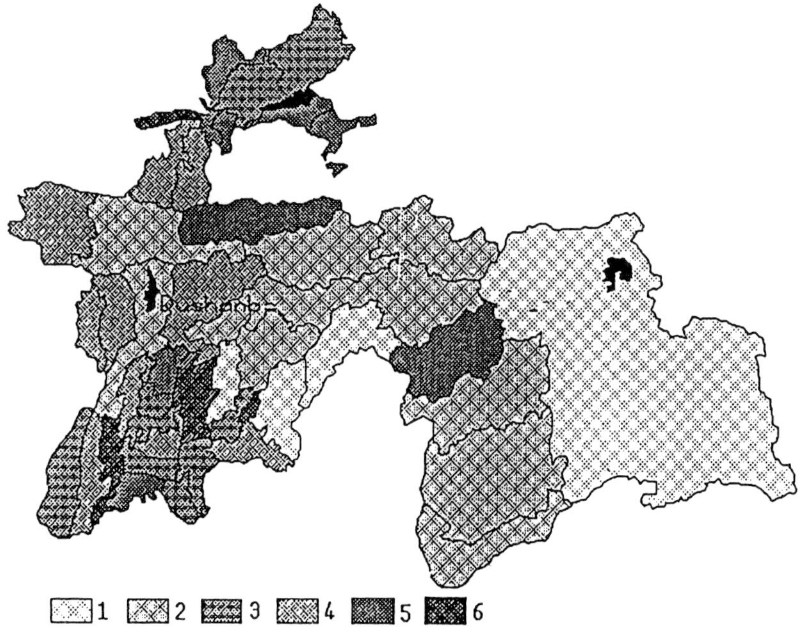

Figure 9.8 Tajikistan: irrigated land (per cent): 1 = 0 – 3; 2 = 3 – 10; 3 = 10 – 20; 4 = 20 – 30; 5 = 30 – 40; 6 = > 40 (from Koshkariov et al., 1994, with permission)

of anthropogenic impacts and as inputs in land-use planning, the focus of the 1991 conference. Human activities on which research has been undertaken include agriculture, the construction of power stations, forestry and recreation. This research has often also involved natural scientists. Yet, unfortunately, no synthesis of the large volume of scientific knowledge deriving from MAB-6 research in Kyrgyzstan has been undertaken, or is planned.

In comparison to Tajikistan and Kyrgyzstan, much less work has been done in the mountains of Uzbekistan, Kazakhstan and Turkmenistan. However, a book summarizing many aspects of the mountains of Uzbekistan, deriving at least partially from MAB-6 activities, has been published (Jumayev, 1989). This describes the natural resources of these mountains and their various economic uses – agriculture, mining, recreation and tourism – and discusses problems of socio-economic development. MAB-6 research in Kazakhstan has concentrated on forests, avalanches and, especially, various aspects of soils: including their composition, erosion, water regime and biota. A few studies have considered geobotany and land use in the mountains of Turkmenistan.

Siberia and the Far East

MAB-6 studies in the many ranges of mountains in Siberia and the Far East have primarily been conducted by natural scientists in the fields of botany (in both tundra and forest ecosystems), geomorphology, glaciology, pedology and zoology. However, a few investigations have been more interdisciplinary, considering environmental impact assessment, medical geography and nature protection. The only MAB meeting to have specifically brought together MAB-6 scientists from Siberia took place in Irkutsk, in March 1988. Nearly all of the twenty-eight participants were from eastern Siberia. Within the general theme of the influence of human activities on mountain ecosystems, the meeting considered four main topics:

(1) Approaches to mapping current and likely future anthropogenic changes in Siberia and the Far East;

(2) Anthropogenic changes in the vegetation of the mountains south of Lake Baikal (Chamar-Daban range);

(3) Stability of the cryogenic landscapes of Yakutia in response to industrial activities; and

(4) Revegetation of land after drainage in the valleys of the Magadan region (Far East).

To date, there has been no summary or synthesis of MAB-6 activities in Siberia and the Far East.

Urals

The only meeting which specifically considered MAB-6 research in the Urals, in relation to the general topic of rational land use in mountain ecosystems, took place in June 1986. A total of sixty-five scientists (botanists, ecologists, foresters, pedologists and zoologists) from the European part of the USSR participated. As the Urals extend across about 15° of latitude, from the industrialized south almost to the Arctic Circle, MAB-6 activities in each of the different parts have had rather different emphases. However, the establishment of reserves to protect endangered or rare ecosystems and species populations has been an important focus of many studies in all parts of the range.

In the southern Urals, research has focused on interactions of human activities – particularly agriculture, industrial pollution and recreation – with meadow vegetation and animals. Both impacts and recovery have been studied. In the central boreal portion, research has concentrated on the formation and development of forest ecosystems, particularly in response to human activities. Work in the northern Urals has included both basic research on mountain meadow flora and fauna (especially rodents) and more applied research on the impacts of human activities on the populations and productivity of these species, many of which are endemic. To date, no synthesis of MAB-6 research in the Urals has been prepared.

CONCLUSIONS

For the first decade of the MAB programme, it does not appear to have had any great influence on research in the mountains of the former Soviet Union. A few studies were labelled as 'MAB-6', but these were generally ongoing, and showed little sign of the interdisciplinary approach of MAB. However, from 1983, MAB-6 activities increased in number and have shown an increasing trend towards interdisciplinary approaches, particularly in the Caucasus, but also – though to a lesser extent – in the Carpathians, Central Asia, Crimea and Urals. Contacts between Soviet scientists and those from other countries have been a crucial impetus to these new approaches. These contacts began with the binational field seminars in the Caucasus, and continued with exposure to the concepts of the Swiss MAB-6 programme and meetings involving international scientists, particularly the 1989 conference in Tsahkadzor, Armenia. The forthcoming book of papers deriving from this conference will be an important addition to the literature of MAB-6.

Since the establishment of the six regional working groups in 1985, research within each region has exhibited different emphases. Except in the Caucasus and Central Asia, the majority of work has been undertaken by natural scientists, though often on applied themes such as land use and natural hazards. However, it should be noted that neither the centralized sectoral system for resource planning and management nor the structure of research institutions in the USSR was particularly conducive to integrated research. In addition, the lack of specific funds for MAB activities doubtless limited their expansion, and the location of the MAB Secretariat within an institute of biology probably limited the involvement of social scientists. Nevertheless, the studies conducted in the Caucasus can be said to have been of the type proposed by the MAB-6 working group in 1973, as shown by both research approaches and resulting publications. To some extent, this is also true for work in other areas, particularly Kyrgyzstan and Tajikistan, but a realistic assessment of the success of MAB is generally limited by a lack of synthesis reports. Again, this partly derives from a lack of funds, a worsening problem for scientific research in all former Soviet republics.

In spite of the demise of the Soviet Union, MAB-6 activities continue in most of the mountain areas of its former republics. However, many activities have ceased, or are in jeopardy, because of the economic and political instability affecting all parts of the former USSR. In spite of minimal funding, the MAB-6 Secretariat in the International Mountain Information and Research Laboratory continues as a focal point for information, particularly through the distribution of a Russian-language version of the World Mountain Newsletter to 300 recipients in educational, scientific and government institutions. It is to be hoped that the current state of instability – which involves conflicts in many mountain regions – will not continue, and that the Eurasian Mountain Association proposed at the October 1992 conference in Vladikavkaz will eventually lead to a new structure for co-operation that is at least as successful as that which existed from 1985 to 1990.

CHAPTER 10

CZECH REPUBLIC AND SLOVAKIA

MOUNTAIN REGIONS

With the principal exception of the Danubian plain in the south-west, much of Slovakia is covered by the hills and mountains of the northern Carpathians (Figure 10.1). Furthest north is the massif of the Western and High Tatras, which includes the highest peak of the Carpathians (Gerlachovsky štit, 2,655 m). South and east of the High Tatras are the Low Tatras (Dumbier, 2,043 m) and the eastern Carpathians (highest peak in Slovakia: Kremenec, 1,221 m). Together, these ranges curve around the

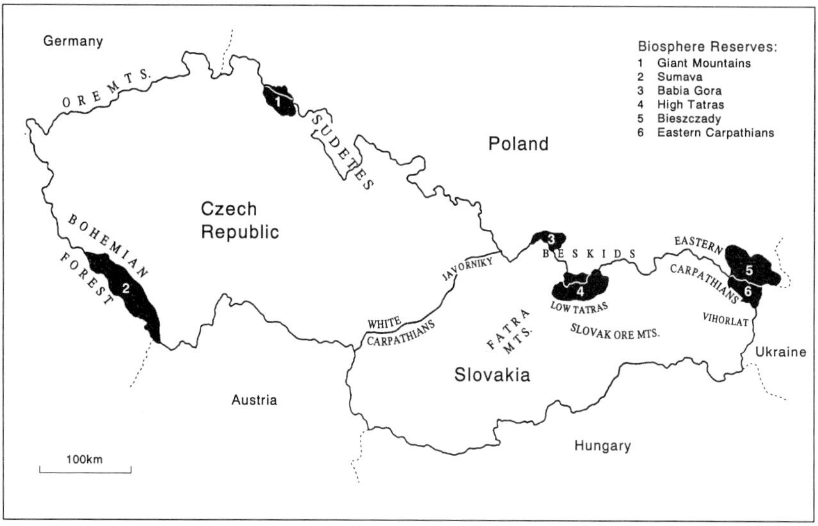

Figure 10.1 Czech Republic, Slovakia and Poland: mountain ranges and biosphere reserves

northern (Polish) and eastern (Ukrainian) borders of Slovakia. A number of lower ranges (High and Low Fatras, Slovak Ore mountains, Vihorlat, etc.), cover much of the remainder of the country, and the White Carpathians and Javorniky range form most of the western border with the Czech Republic. The other mountains of the Czech Republic are of Hercynian age, and occupy most of the country's borders. On the Polish border are the Sudetes, whose highest range is the Giant Mountains (*Krkonoše*; highest peak: Snězhka, 1,602 m). Along the German border are the mountains of the Bohemian Forest (highest peak in the Czech Republic: Roklan, 1,453 m) and the Ore Mountains (Klinovec, 1,244 m).

The mountains of the Czech Republic and Slovakia are largely forested; few have a significant proportion above the timberline. The principal exceptions are the High Tatras and some of the other Carpathian ranges and, to a lesser extent, the Giant Mountains and the Hruby Jesenik to their east; all of the highest areas are within National Parks or Protected Landscape Areas. Many of these on borders are adjacent to a protected area in the neighbouring country. However, many of these ranges, particularly those on the Polish and German borders, are subject to severe air pollution – a critical issue both for protected areas and for those where forestry remains the dominant activity. In addition, the development of recreational and tourist infrastructures has been substantial, especially in the Giant and Tatra mountains, which provided the only reliable sites for skiing which were accessible to citizens of socialist countries during the Communist era.

ORGANIZATION OF MAB ACTIVITIES IN MOUNTAIN AREAS

For most of the period under review, the Czech Republic and Slovakia constituted one country: Czechoslovakia. Consequently, most of this section and the conclusions consider MAB within all the mountains of the former Czechoslovakia.

When a National Committee for MAB was established in Czechoslovakia in 1975, responsibility for the fourteen Project Areas was evenly divided between Czech and Slovak scientists. Unfortunately, the MAB-6 subcommittee was never convened by any of the three scientists who successively held the post of chairman (Czechoslovak National Committee, 1990). However, considerable work in mountain areas was associated with MAB within the scope of both Project 2 (temperate forests) and Project 8 (biosphere reserves). Sub-committees for both of these projects have been active since the mid-1970s and, notably, the Institute of Forest Ecology in Brno (Czech Republic) has been UNESCO's Information Centre for Temperate Forests of the Northern Hemisphere (i.e. MAB-2) since 1976. In addition, one should note that a number of major international scientific meetings, both officially within the scope of MAB and involving scientists active in MAB, have taken place in the mountains of the Czech Republic and Slovakia since the 1970s.

In 1990, the National Committee decided to constitute a number of working groups on 'pragmatic' themes which did not necessarily match MAB's original project areas. One of these groups, with a Czech chairman and a Slovak vice-chair, considered mountain ecosystems. Members of this group proposed networks for protected areas in the Carpathians and for research in the Hercynian mountains at the 1991 EuroMAB meeting. The former has been implemented to some extent, with a focal point in the Tatra National Park in Slovakia. In the Czech Republic, a first activity of the latter was the development of joint projects with scientists involved in MAB-6 activities in Poland. One of these, a 'Sudety Mountains Biodiversity Protection Program,' was approved in 1992 for funding within the 'Polish Forest Biodiversity Protection Project' of the World Bank's Global Environment Facility. In contrast to the lack of scientific trans-border co-operation during the communist era, this programme has a bi-national scientific committee. Another project, with a greater emphasis on ecosystems above treeline, has been proposed to the Council of the European Communities by a consortium of Czech, German and Polish scientists.

After the division of Czechoslovakia into two states at the beginning of 1993, separate national MAB Committees were constituted in each country. In September 1993, the Czech National MAB Committee, together with many other Czech organizations and the IUCN, sponsored an international conference in Krkonoše National Park on 'Mountain national parks and biosphere reserves: monitoring and management', with 100 participants from across Europe as well as North America. The meeting endorsed the establishment of a 'Centre for the Management and Restoration of Disturbed Ecosystems in Hercynian Mountains' (CEHEM) with an initial base in the Krkonoše National Park. However, links to scientists and resource managers in other Hercynian mountains (e.g. Massif Central, Vosges, Black Forest) have yet to be formalized.

SCIENTIFIC RESEARCH AND BIOSPHERE RESERVES

Scientific research in the mountains of the Czech Republic and Slovakia has a very long history, dating back to the seventeenth century in both the Tatra and Giant Mountains. Both the Krkonoše and Tatra National Parks have published annual collections of scientific papers for a number of years. These two parks, and also Šumava, produce popularized publications describing scientific research and other issues of interest.

In general, it appears that, although projects directly under the auspices of

MAB-6 were not undertaken in the mountains of Czechoslovakia during the 1970s and 1980s, documents deriving from the MAB Secretariat were important in introducing new concepts, particularly those of biosphere reserves and multidisciplinary research including both natural and social scientists. Apart from the lack of an official MAB-6 committee to develop projects in specific mountain areas, another reason for the lack of 'official' MAB-6 projects was that the two major mountain National Parks (Krkonoše (Czech Republic) and High Tatras (Slovakia)) have considerable numbers of permanent scientists. Both these and other scientists, particularly from the Czechoslovak and Slovak Academies of Sciences, had been undertaking scientific research for many years when the MAB programme started, so that MAB-6 was not needed as a stimulus to research, as it was in mountain regions in other countries. Also, as mentioned in the introduction, the 5-year plans for scientific research were a definite limitation to developing interdisciplinary programmes, which tend to evolve during their dynamic history. Finally, access to some mountain areas, especially those along the German border, was limited for 'security' reasons.

None of the first four biosphere reserves designated in Czechoslovakia was in a mountain area: another contrast with other east and central European countries. This was a conscious decision of the nominating committee which felt that, given constraints on staffing for scientific research and management, less attention should be given to areas which were already adequately protected in other ways (i.e. the National Parks of the Tatras (Slovakia), Krkonoše and Šumava (Czech Republic)) (Jeník, 1987). The High Tatras and Krkonoše National Parks were designated before the MAB programme, in 1949 and 1963, respectively. Šumava was declared a Protected Landscape Area (PLA) in 1963, and nearly half of this was designated a National Park in 1991, a year after the whole PLA had been designated a biosphere reserve.

In the 1990s, the additional value of international recognition for the high levels of anthropogenic pressure on the important ecosystems of Krkonoše and the High Tatras led to proposals for the designation of bilateral biosphere reserves with their Polish counterparts. These were approved by the MAB-ICC in 1993. Another proposal for a trilateral Polish–Slovak–Ukrainian biosphere reserve in the eastern Carpathians (including the Slovak PLA designated in 1977) was only partially approved; the Ukrainian portion has not been designated for the time being. An English-language book on the biosphere reserves of the Czech Republic and Slovakia was published in 1994 (Jeník and Price, 1994). The locations of the current mountain biosphere reserves are shown in Figure 10.1.

In spite of the lack of 'official' MAB-6 projects in the mountains of the Czech Republic and Slovakia, and although the history of biosphere reserves in these mountains is rather short, the remainder of this review focuses on the scientific research that has been conducted in the four mountain National Parks/biosphere reserves since 1975. This is an appropriate focus for the review because, as discussed above, MAB concepts are recognized as a conceptual stimulus to research during this period; and also because future activities in biosphere reserves and the proposed networks on Hercynian and Carpathian mountains will build on this work.

Giant Mountains (Czech Republic)

The Giant Mountains form the highest part of the border between the Czech Republic and Poland. Their location and height are associated with a number of remarkable characteristics, both positive and negative. To consider first the positive aspects, the mountains have a very diverse biota. This includes almost half of the native plant species of the Czech Republic, from subalpine forests to alpine tundra: the only example in the nation. The fauna include many glacial relicts and species at their northern limit of distribution (Figure 10.2). Finally, the mountains also have a reliable and regular snow cover and steep, glaciated scenery.

Unfortunately, these mountains are affected by two major types of anthropogenic stresses. Air pollution from Germany, the Czech Republic and Poland has greatly affected much of the vegetation, most obviously in the highest spruce forests, many of which are artificial. Reforestation began centuries ago in connection with glass-making and mining. As well as direct damage to trees by acid precipitation, they have been severely affected by windstorms and insect epidemics (Figure 10.3). The other major impacts derive from tourism, which began in the eighteenth century. About 7 million people visited the Czech part of the mountains each year in the late 1980s, though visitation has decreased considerably in recent years. Hiking has resulted in erosion and vegetation changes along heavily used paths. Other negative impacts are associated with mountain chalets, recreation centres and sports facilities, including ski areas. In 1984, the World Conservation Union (IUCN) recognized the Krkonoše National Park as one of the twelve most endangered in the world. The park comprises the core zone (8,432 ha) and buffer zone (27,925 ha) of the biosphere reserve. The transition zone, where most of the population of 26,700 live, is 18,430 ha in area (Figure 10.4).

In relation to their size, the Giant Mountains must be one of the most-researched mountain ranges in the world, with a history of published scientific research dating back to 1607. A number of syntheses of scientific work have been published, concerning both a wide range of scientific disciplines (e.g. Sykora et al., 1983) and individual disciplines (e.g. a flora describing 1,149 species: Sourek, 1969). During the review period, an

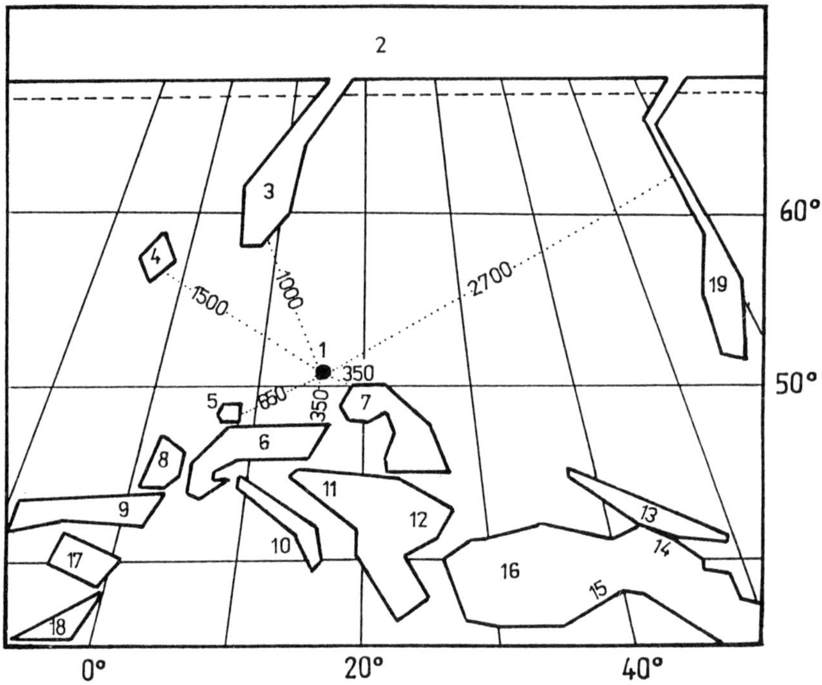

Figure 10.2 Location of the Giant Mountains (1) in Europe, showing distances (in km) to areas with corresponding arctic, alpine, and subalpine systems in the Arctic (2), Scandinavia (3), Scotland (4), Vosges and Black Forest (5), Alps (6), Carpathians (7), Massif Central (8), Pyrenees and Cantabrian Mountains (9), Apennines (10), Dinarids (11), Balkans (12), Caucasus (13), Trans-Caucasus (14), Taurus (15), Anatolia (16), Central Sierras (17), Sierra Nevada (18) and Urals (19) (from Jeník and Price, 1994, with permission)

essential source of scientific documents is the annual collection of scientific papers, *Opera Concortica*, published since 1964, with English and Russian summaries. Bibliographies of scientific works have been published within this series (e.g. Dvořak and Pilous, 1988), and also as separate documents. These show the substantial amount of scientific work conducted in this area. For instance, a selected bibliography of publications from 1963 to 1978, for both the Czech and Polish Giant Mountains, includes 3,627 references (Belochova, 1982). This is an exception to the general rule of minimal communication between scientists on the Czech and Polish sides of the border; for instance, Polish scientists have published very few papers in *Opera Concortica*. In 1992, the Administration of Krkonoše National Park began publication of a yearbook (with English summaries) which summarizes conservation, research, resource management, information/documentation and other activities.

Figure 10.3 Czech Giant Mountains from the alpine zone looking north, showing forests logged to remove trees damaged and killed by acid precipitation

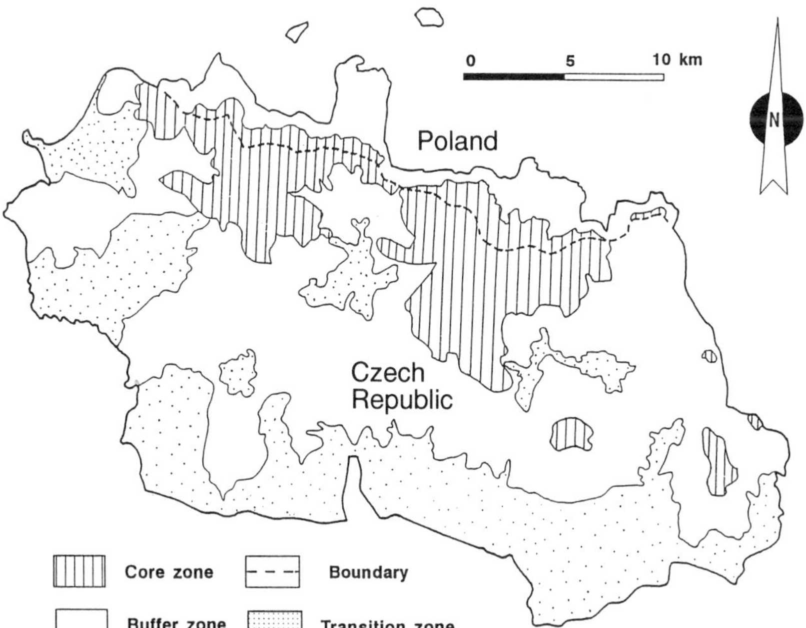

Figure 10.4 Zonation of the International Giant Mountain biosphere reserve (from Jeník and Price, 1994, with permission)

The majority of scientific research conducted in the Giant Mountains has been undertaken by natural and physical scientists, with examples from all disciplines and a wide variety of subdisciplines. A smaller proportion of work has considered subjects such as nature protection, history, ethnography and tourism. One example of the predominance of natural science is provided by the bibliography in volume 25 of *Opera Concortica* (Dvořak and Pilous, 1988). This shows that less than 10 per cent of papers – published by scientists from both the National Park and the Academy of Sciences – from 1964 to 1988 were on other subjects. However, there is little evidence of multi- or interdisciplinary work, particularly conducted jointly by natural and social scientists. The present emphasis of applied research is the reconstruction of forests damaged by air pollution, with significant funding from the Dutch Forests Absorbing Carbon Dioxide Emissions (FACE) Foundation over a 4-year period (1993–1996). This work includes both Czech and Dutch scientists and foresters.

Šumava (Czech Republic)

The low mountains of Šumava (highest peak in the Czech Republic: Plechy, 1,378 m) adjoin the Bavarian Forest in Germany and the Austrian Mühlviertel. This region contains almost 200,000 ha of continuous forest: the largest forest in central Europe. The average forest cover is particularly high, with many primeval and natural stands as well as secondary forests. Although many of the forests have a long history of human use, during the Communist era Šumava was a 'frontier security zone', in which they were left to regenerate without management. This has resulted in a unique pattern of ecosystems, from forests to significant areas of wetland ecosystems: the largest concentration in the Czech Republic. There are many rare and endangered plant and animal species, mainly in the core zone (42,224 ha) of the biosphere reserve. The buffer zone has an area of 68,893 ha, and the transition zone, which has a population of 22,900, is 56,000 ha in size (Figure 10.5).

On the German side of the border, research has been ongoing for many years in the Bayerischer Wald National Park. In contrast, post-war scientific research on the Czech side, including the Šumava area, was limited until recently by 'security' considerations and the consequent constraints of military administrative requirements; and also by lack of co-operation between local administrative authorities until the creation of the National Park in 1991. Consequently, research was restricted to studies conducted by individual scientists who were willing to fulfil the military requirements and by other scientists who undertook routine hydrological and climatological monitoring. Independent research primarily focused on mires and

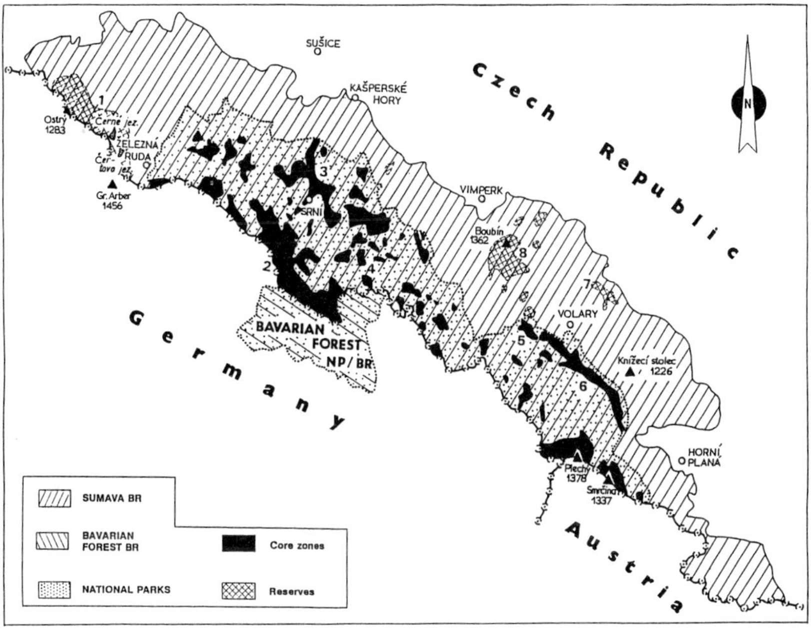

Figure 10.5 Zonation of the Šumava biosphere reserve and adjacent protected areas (from Jeník and Price, 1994, with permission)

forests, both managed and natural (especially the Boubin primeval forest, a nature reserve since 1858). Since the creation of the biosphere reserve, the amount of scientific research has been increasing, but there are very few permanent scientists.

High Tatras (Slovakia)

The High Tatra massif is the central and highest part of the Tatra National Park (TANAP), which also includes the West and Belianske Tatras. The dominant rocks in the three ranges are, respectively, granitic igneous, metamorphic and limestone. The geological diversity is reflected in very different landscapes, with high alpine peaks – seventeen above 2,500 m – and glaciated valleys in the High Tatras; limestone cliffs in the Belianske Tatras; and karst features, including many caves and sinkholes, in the West and Belianske Tatras. From a biogeographical viewpoint, the Tatras are notable as an island including rare montane and alpine ecosystems with a unique range of plant and animal species. As the highest mountains between the Caucasus and both the Alps and the North Pole, they are a major

Figure 10.6 Part of the High Tatras of Slovakia from the air, showing the resort of Strbske Pleso

cross-roads on both north–south and east–west migration routes of alpine and arctic species.

The most long-term human use of the area was sheep-grazing, which began in the fourteenth century and continued until 1955. Commercial forestry began in the sixteenth century and continues today on a limited scale. As a result of these activities, the treeline was lowered by an average of more than 200 m. However, the altitude, steep and dissected relief, harsh climate, unsuitable soils and problems of accessibility that characterize the Tatras meant that they remained largely uninhabited until the second half of the nineteenth century, when a number of settlements were built for recuperation, recreation and tourism. Nevertheless, the traditional economy, based on pastoral agriculture and forestry, continued until the Communist era, when the mountain forests and pastures owned by the villages of the foothills were nationalized, and then incorporated into TANAP in 1949. The economy of the area has since been based on tourism, with about 6 million visitors a year in the late 1980s (Figure 10.6). In the 1990s, visitation has decreased very substantially to perhaps only half this number; the future of tourism is increasingly uncertain. The core zone (49,633 ha) and the buffer zone (23,744 ha) of the biosphere reserve are in TANAP. The transition zone (39,844 ha) includes all of the tourist settlements.

Scientific research in the High Tatras has a long history, dating back to the early eighteenth century. Some of this has been summarized in publications relating to specific disciplines, often including research from both the Slovak and Polish sides (e.g. Koncek, 1974; Pawlowski, 1956). While there is no overall synthesis of scientific work undertaken in the Slovak Tatras, many of the volumes of the annual compilation *Zbornik Prac o Tatranskom Narodnom Parku* (Treatises concerning the Tatra National Park) have specific themes, including botany, zoology, meteorology and climatology and anthropogenic influences on ecosystems. This series, which includes English, German and Russian summaries, has been published since 1957. It has also included a number of bibliographies, of which the most complete was published on the 35th anniversary of the National Park (Marcek, 1984).

During the review period, research in TANAP has been conducted both by members of the Park's research station (fifteen full-time scientists) and also by other scientists, particularly from the Slovak Academy of Sciences and Slovak universities, who have published their findings in a wide range of publications. This research has primarily considered natural and biological sciences, although research in archaeology, ethnography, nature protection and recreation has also taken place. There has been little, if any, interdisciplinary work.

In 1992, work began toward a major 5-year research programme (Project MONTAN) focusing on the assessment and management of anthropogenic stresses. This is planned to be multidisciplinary, with integrated research both at the scale of TANAP as a whole and for four representative watersheds. Again, this applied project will be primarily conducted by natural scientists, although some work will consider patterns of recreation and their impacts. However, few of the planned activities in this project have begun, because of financial and planning problems. Finally, co-operative research and monitoring activities have been undertaken since 1985 with the adjacent Polish Tatrzański Park Narodowy, focusing on acid precipitation and its effects on forests. It is anticipated that such co-operative work will increase within the context of the bilateral biosphere reserve.

Eastern Carpathians (Slovakia)

The Eastern Carpathians biosphere reserve is at the junction of Slovakia, Poland and Ukraine. These low mountains are in a biogeographically important location, both as an outlying island of East Carpathian ecosystems adjacent to the West Carpathian ecosystems prevalent in Slovakia and, more generally, as a region with a mixture of East and West Carpathian species. While the area is mainly covered by beech forests, of which a relatively high proportion have never been managed, the most unusual ecosystems are alpine grasslands (*poloninas*). Most of these are secondary communities which developed as a result of cattle-grazing, which began with the first phase of settlement in the fourteenth century. The local economy is based on forestry, with some agriculture as well as hunting and fishing, the main activities for which tourists visit this remote area. The Slovak part of the biosphere reserve has a population of 3,700, who live in the transition zone (23,585 ha) of the reserve. The core zone (seven components) and buffer zone have areas of 2,643 ha and 14,373 ha, respectively.

Scientific research has taken place intermittently in this area since the early nineteenth century. In the inter-war period, a major multidisciplinary project was conducted by Czechoslovak scientists in the Stuzhica valley, which is now in Ukraine. After the Second World War, silvicultural research was conducted in the forests. Other work in both forests and areas above timberline was conducted by individual scientists, from a number of disciplines. A major summary of scientific information for the area has been prepared by Hadac and Terray (1991).

Other mountain areas

In addition to the research described above in the four mountain National Parks/biosphere reserves, some work has been conducted in lower-altitude mountain areas within the scope of MAB, particularly in forest areas within the framework of MAB-2. In the Little Carpathians of Slovakia (maximum altitude: 768 m), members of the Slovak Academy of Sciences conducted a project on soil ecology from the mid-1970s to the mid-1980s. Most of the published report is rather narrow in scope: following two papers on the ecology and forest types are thirteen papers on soil fauna (Nosek, 1986). There is no synthesis. In 1989, representatives of the Czechoslovak and Ukrainian MAB Committees agreed to develop a collaborative programme on the impact of human activities on forest ecosystems of the Carpathian mountains (MAB Newsletter for Project 2, No. 7). This has not proceeded very far, for a variety of institutional and economic reasons.

Since the mid-1970s, research has also been conducted in the lower mountains of the Czech Republic, by scientists from the Faculty of Forestry at the Agricultural University of Brno. This work has considered the ecological consequences of mechanization and clear-cutting in spruce monocultures in the Bohemian-Moravian uplands, and the ecological impacts of acid precipitation – and its amelioraton by liming – on forests in the Beskid mountains (Klimo, 1981, 1990; Prax and Raev, 1985).

CONCLUSIONS

Despite the lack of a working structure for the organization of MAB activities in the mountain areas of the Czech Republic and Slovakia until 1990, a considerable amount of scientific research has been conducted in these areas which has been influenced to some extent by the concepts of MAB. The two predominant frameworks for this research have been programmes in the mountain National Parks (particularly Krkonoše and High Tatras) mainly undertaken by staff scientists and others from the Academies of Sciences, and MAB-2 projects, which have generally concerned lower mountain areas. However, MAB and MAB-related projects in the mountains of the Czech Republic and Slovakia have generally shown few signs of the multi- and interdisciplinary approaches endorsed by MAB, and their emphasis has been overwhelmingly on the natural sciences. Given the 5-year planning approach for science until 1989, it seems fair to say that most projects would have happened in any case, but benefited to some extent from the international connections provided by MAB.

Considering the general lack of international co-operation until recently – for instance, similar projects were often conducted on both sides of the Czechoslovak – Polish border, but with no interactions – the reorganization of MAB for mountain research in 1990, focusing on the new mountain biosphere reserves and linked to the development of new networks for the Carpathian and Hercynian mountains, was a promising sign. However, these new approaches could not solve a critical problem that has characterized MAB in the Czech Republic and Slovakia since its beginnings: a lack of specific funding for MAB projects. During the Communist era, the Academy of Sciences provided a part-time position for the secretary of the MAB National Committee from 1979 to 1987, when this post became full-time, as it remains; though the secretary is now responsible for the Czech Committee only. However, this has been the only direct funding for MAB, other than occasional support for travel and conferences.

The proposed joint research projects in the Giant Mountains involving Czech, Polish, German and Dutch scientists, as well as other research and monitoring programmes in biosphere reserves, are a pragmatic approach to undertaking research in times of scarce national funding. It is to be hoped that the designation of trans-boundary biosphere reserves will be a stimulus to co-operative research once the present administrative problems have been resolved. Nevertheless, with the demise of the Czechoslovak MAB National Committee and biosphere reserve committee – two of very few national scientific committees with roughly equal participation of scientists from the Czech and Slovak Republics – and the constitution of National Committees for the Czech Republic and Slovakia, the future of MAB mountain activities in the two countries is hard to predict.

CHAPTER 11

POLAND

MOUNTAIN REGIONS

All of Poland's mountains are in the southern part of the country; they form the majority of the border with the Czech Republic and Slovakia, from the western border with Germany to the eastern border with Ukraine (see Figure 10.1). The western mountains are part of the Hercynian Sudeten mountains. From west to east, these are the Gizerskie mountains (highest peak: Smrek, 1,124 m), the Giant Mountains (*Karkonosze*: Sniezka, 1,603 m), and the lower ranges surrounding the Klodzko valley. The eastern ranges are the northernmost extension of the Carpathians. The majority of these are the Beskids, part of the outer Carpathians. These begin to the west of the Tatras, pass to their north, and continue eastwards into the Ukraine. The highest peak of the Beskids, Diablaka (1,725 m) is in the Babia Góra area west of the inner Carpathian Tatras, which include Poland's highest peak (Rysy, 2,499 m). Among the Beskid ranges are the Gorce (Turbacz, 1,310 m) and Pieniny (Wysoka, 1,052 m), respectively north and east of the Tatras. The easternmost part of the Polish Beskids are the Bieszczady mountains (Tarnica, 1,346 m).

Few of Poland's mountains have a significant proportion of their area above timberline. The Polish Tatras, which are only a small part of the range, have a distinct alpine zone. The Polish Giant Mountains also include only a relatively small part of the range although, with steep north-facing cirques, this is markedly different from the Czech side. Along the Beskid ranges, few summits reach timberline. The Bieszczady are unusual in having no subalpine coniferous forest zone; beech forests abut directly on subalpine scrub and grassland communities.

Anthropogenic stresses are an important focus of concern and research in Poland's mountains (Grodzinska and Olaczek, 1985). All are influenced to some degree by acid precipitation, though this tends to decrease from west to east. Tourism is another important anthropogenic stress in limited areas, especially the Tatras and Giant Mountains, although numbers of tourists have decreased markedly in the 1990s. The highest parts of these two ranges, and also those of the Babia Góra, Bieszczady, Gorce and Pieniny ranges, are protected as National Parks. All of those along the national border abut National Parks or Protected Landscape Areas in the Czech Republic and Slovakia; and, in the case of the Bieszczady, also a nature reserve in Ukraine.

ORGANIZATION OF MAB ACTIVITIES IN MOUNTAIN AREAS

A MAB National Committee was formed in Poland in 1973, under the auspices of the Committee for Man and Environment of the Polish Academy of Sciences (PAS). The MAB National Committee was combined with the existing National Committee for SCOPE. Responsibility for MAB-6 projects was given to Prof. K. Zabierowski, Director of the Nature Protection Centre of the PAS, who initiated six MAB-6 projects, which were listed in the MAB Secretariat's 1981 compilation of MAB projects (see below). Another MAB-6 activity in the late 1970s, beginning in 1978, was the compilation of various syntheses of scientific information on various themes for both the Polish Carpathians and the Carpathians as a whole. However, these were not published; the compilation on the botany of the Carpathians is currently in press.

In 1979, Prof. Zabierowski left the Nature Protection Centre but, as chairman of a committee on the management of mountain soils, remained responsible for MAB-6. In meetings of the National Committee during the 1980s, interest was expressed in conducting research, within the scope of MAB-6, relating to mountain agriculture and forestry. However, there is no evidence of MAB-6 activities or publications during this period. In 1992, the National Committee reconstituted a subcommittee for MAB-6. This is presently developing research projects, primarily in mountain biosphere reserves, and also organized the international conference on ecological zonation in mountains, which took place after the fourth EuroMAB meeting in Zakopane in September 1993.

SCIENTIFIC RESEARCH AND BIOSPHERE RESERVES

Scientific research in the Polish mountains dates back at least to the seventeenth century. Changes in the country's borders pose some problems in tracing the history of research in some areas, particularly the Giant and Bieszczady mountains. In recent decades, most research in mountain areas has been conducted by scientists from the Polish Academy of Sciences, universities, and research institutes. The oldest National Park research station, in the Tatra National Park, was founded in 1980, and another was established in Bieszczady National Park in 1991. Other National Parks have only one or two permanent scientists.

In 1981, the Polish National Committee published a bibliography of 80 publications which appeared from 1960 to 1978 and were regarded as contributing to MAB-6 (Polish National Committee, 1981). The MAB Secretariat's 1981 compilation of research activities within the scope of MAB-6 in Poland also lists seven projects (UNESCO, 1981). All of these were organized, in various parts of the Carpathians, by scientists in Krakow. Six projects concentrated on the impacts of tourism, reservoirs and other land uses on mountain environments; all were 4- to 6-year projects starting in 1976 or 1977. A 4-year project, directed by Prof. Zabierowski, on agricultural models was to start in 1981. However, while papers on their topics were published, these MAB-6 projects do not appear to have resulted in publications that identify their origins.

Although there has been a lack of 'official' MAB-6 projects, much scientific research was undertaken during the period of the MAB programme in mountain regions. A number of documents have been published which bring together significant amounts of information about the various Carpathian ranges with regard to nature protection (Klimek, 1989) and tourism (Anon, 1989; Zabierowski, 1982). A particular focus for research has been the areas which are now biosphere reserves. Only one of Poland's first four biosphere reserves, created in 1976 and 1977, was in a mountain area: Babia Góra. However, three more were designated in 1992 (see Figure 10.1; Breymeyer et al., 1994). All are joint biosphere reserves, with the Czech Republic (Giant Mountains) or Slovakia (Tatras, Bieszczady). As noted in the previous chapter, a trilateral international park and biosphere reserve in the latter area, also including the Stuzica reserve in the Ukraine, is under consideration. The following sections briefly describe research conducted in these areas during the course of the MAB programme. This research forms an essential basis for future research in, and management of, these biosphere reserves.

Babia Góra

Babia Góra includes the highest peak in the Beskids, which has a small alpine zone. The north slopes are generally steep and glaciated, while the south slopes are more gentle. Most of the area of Babia Góra National Park is covered by forests of many types, from beeches at the lowest altitudes to dwarf pines in the subalpine zone. Many of these forests have experienced little or no management. However, the proportion of spruce forest is greater than the natural state as a result of management in past centuries, and grazing by sheep and cattle has significantly changed the structure of grassland and understorey plant communities. Since the declaration of the National Park (1,734ha) in 1954, these forestry and pastoral activities have ceased. The National Park was approved as a biosphere reserve in 1977.

Scientific research in Babia Góra has been conducted since 1804, and the considerable volume of information available by the early 1980s was published in 1983 (Zabierowski, 1983). In addition to chapters on all of the major natural and biological science disciplines, this synthesis includes chapters on agriculture, forestry, hunting, tourism, the cultural heritage policy and education. During the 1980s, forest areas were the primary focus of research, which emphasized forest succession and age structure. Another focus has been on the precipitation and accumulation of heavy metals in soils and vegetation. Research on other diverse topics, including phytosociology, entomology and tourism, has also been undertaken.

Bieszczady

As mentioned above, the Bieszczady mountains are unusual in that a subalpine coniferous forest zone does not separate beech forests and the subalpine grasslands (*poloniny*). This situation derives at least partly from the area's long history of human use, which dates back to the fifteenth century when the Bojkowie and Lemkovie tribes colonized the mountains, which they used for pasturing animals. The descendants of these tribes were expelled from the area during and after the Second World War. Since the 1950s, the area has been partly resettled (Jeník, 1983). Most of the Bieszczady is occupied by beech forests, which include many virgin stands in spite of logging since the fifteenth century for construction, fuel and, particularly, charcoal production. In 1973, 5,725ha of the region were protected as a National Park, which was expanded to 27,064ha in 1990. The adjacent San Valley and Cisniensko-Wetlinski Landscape Parks (35,835 and 46,025ha, respectively) were established in 1992. Together, these protected areas form the Polish portion of the Eastern Carpathians biosphere reserve.

Despite the remoteness of the Bieszczady, scientific research has been conducted in the area since the early nineteenth century, and a bibliography of 1,875 publications until 1975 has been published (Partyka, 1977). Research during the 1970s and 1980s was limited; a few studies by individual scientists considered mammal populations, vegetation succession on balds, stream invertebrates and theoretical considerations for tourism. In 1991, a scientific research station with ten staff was established, and is now developing research and monitoring activities.

Figure 11.1 Maly Staw cirque on the northern side of the Giant Mountains biosphere reserve, Poland

Giant Mountains

The Polish side of the Giant Mountains (*Karkonosze*) is similar in many ways to the Czech side (*Krkonoše*) described in the previous chapter. However, because of its north-facing aspect, the Polish side is distinguished by steep glacial cirques, many containing lakes (Figure 11.1). Both air pollution and tourism – about 2.5 million visitors a year in the late 1980s – are major management concerns in the forests, which are mostly spruce. Though spruce is the natural dominant species at higher elevations, those at lower elevations were planted in the nineteenth century, largely replacing the indigenous beech. The area has a history of human use (pastoralism and forestry) extending back to the fifteenth century. In 1959, a National Park (5,564 ha) was established. This has been divided into a core zone (1,117 ha) and a buffer zone (3,864 ha) within the new biosphere reserve (see Figure 10.4). The boundaries of a transition zone are still under discussion.

Scientific research has been conducted since the eighteenth century in the Polish Giant Mountains which, until the end of the Second World War, were part of Germany. The results of research until the early 1980s have been brought together in a major synthesis (Jahn, 1985). In addition, a bibliography of 3,344 publications from 1945 to 1984 was published in 1989 (Kuzma, 1989) in the series of *Papers of the Karkonosze Scientific Society*. This series began in 1973, and more than fifty volumes have been published, including a recent group of papers from various disciplines on the 30th anniversary of the National Park (Konca and Bugaj, 1991).

The longest continuous series of scientific measurements are climatological: data are available from Snieszka, the highest peak, since 1824, and continuously (for some parameters) since 1881. There are other climatological stations, although the period of record generally ceased in the early 1930s and did not begin again until the 1950s. In addition, air quality is measured at seventeen sites, and more are planned. Forest ecosystems, particularly those above 1,000 m, have been a primary focus of research during the 1980s. The forests have experienced – and continue to experience – massive changes deriving from various combinations of the effects of acid precipitation, windstorms and insect epidemics. However, in contrast to the Czech side, where many damaged forests have been logged, those on the Polish side have been left for natural processes to take their course.

Research on forests has considered succession (especially after damage by acid precipitation), soils, insects and the survival of tree seedlings. One of the main objectives of the entomological research is to develop methods of bioindication of damage to ecosystems by acid precipitation, as insects are particularly rapid indicators of ecological change. With respect to regeneration, an important focus is on the role of mycorrhiza in forest regeneration, particularly for beech, which was replaced by spruce plantations during the nineteenth century, but appears to have a greater resistance to acid precipitation. Other research has considered primary vegetation succession, phytosociology, geology, hydrology and tourism. In the early 1990s, a number of major projects are beginning. One of these, involving about 60 scientists from the PAS and institutions in Wroclaw, considers stressed mountain ecosystems. It has four major components: small mammals, vegetation, air pollution and climate. Another is the joint Czech–German–Polish project on forest biodiversity, funded by the World Bank, described in the previous chapter. At present, there is a gradual increase in co-operative research and monitoring activities with Czech scientists and National Park employees.

Tatras

The Polish part of the Tatras occupies only a small proportion of the range, though it includes the full diversity of bedrock types: granitic igneous, metamorphic and sedimentary (Figure 11.2), in the High, Western and Reglowe portions, respectively. The human history of the area (Figure 11.3) dates back to the fourteenth century, when mining began, reaching its heyday in the eighteenth and nineteenth centuries. Summer sheep-grazing

Figure 11.2 Czerwone Wierchy in the limestone West Tatras (photograph by W. Siarzewski)

began in the fifteenth century. Shepherds burned and cleared forests to provide pastures, lowering the timberline by 100–200 m. The forests were also substantially altered by intensive forestry in the nineteenth and twentieth centuries and, more recently, have been affected by air pollution from both local and distant sources. In the 1840s, the railway from Krakow reached the area, and the village of Zakopane began its development into a major tourist resort (Figure 11.4). It now has 30,000 inhabitants and is visited by about 1.6 million visitors a year, a significant drop from the 3 million in the 1980s. Most tourists visit in summer, but there is also a winter season. The ski area is within the National Park (21,164 ha), which was established in 1954 (Figure 11.5). Since this time, limited forestry has continued in private forests (6,350 ha) within the Park boundaries. Sheep grazing was eliminated by 1970, but began again in 1981 to a limited extent in selected pastures, after recognition of its cultural and ecological value.

Scientific research in the Polish Tatras has a long and rich history, dating back to the end of the eighteenth century. A vast number of scientific studies in diverse disciplines have been undertaken throughout the nineteenth century and up to the present day. In addition to the great volume of knowledge deriving from such studies, long-term data sets are available for vegetation and for climate. A meteorological station was established in the town of Zakopane, which adjoins the National Park, in 1911, and high-

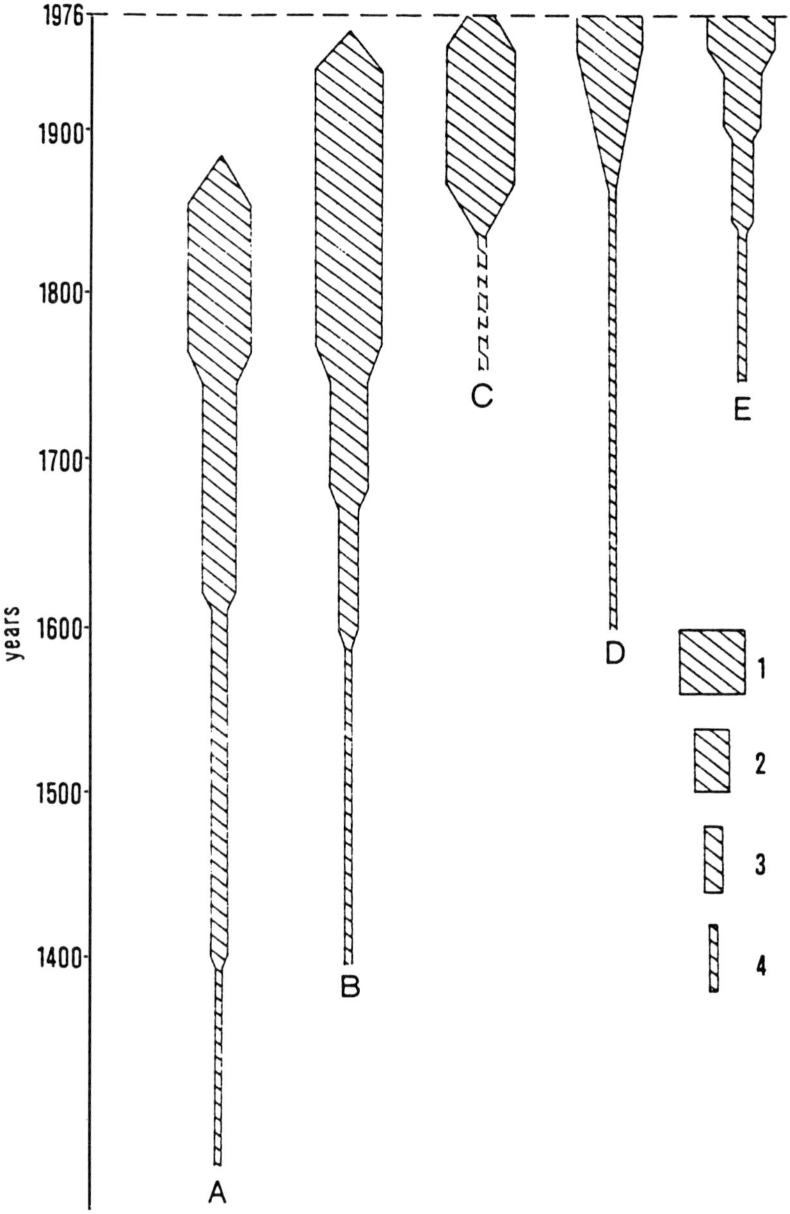

Figure 11.3 The main types of human impact on plant cover in the Polish Tatras, *c.* 1300 to 1976. 1–4 = degrees of intensity of impact (1 = greatest); A = mining and metallurgy; B = sheep and cattle grazing; C = forestry and forest exploitation; D = long-range air pollution; E = tourism (from Mirek, 1992, with permission)

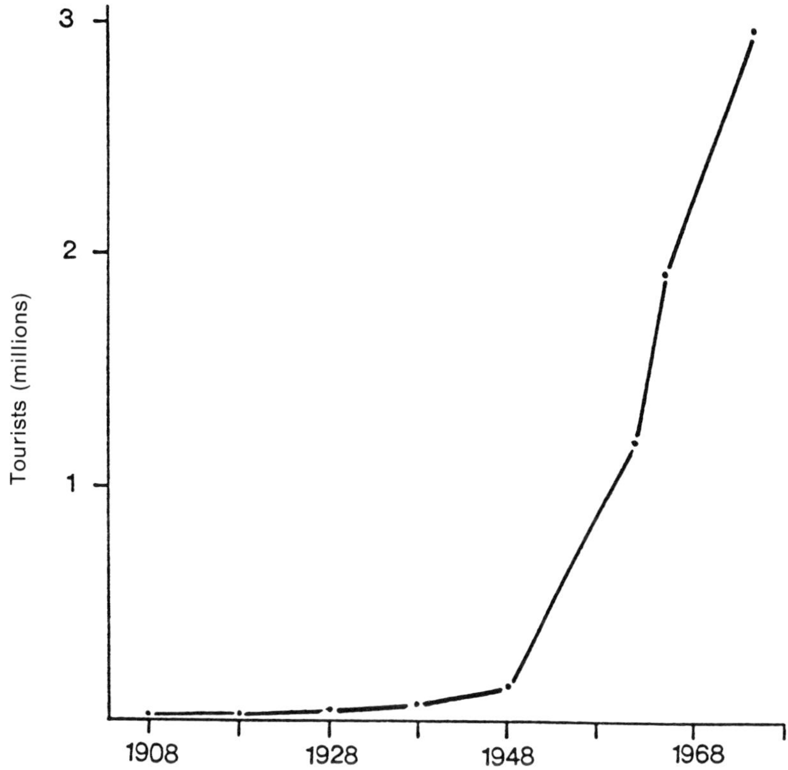

Figure 11.4 Number of tourists (millions) visiting Zakopane and the Tatra National Park 1908 to 1978 (from Mirek, 1992, with permission)

altitude stations were established at Hala Gasienicowa (1,520 m) in 1913, and at Kasprowy Wierch (1,991 m) in 1937. Research has been organized through the local museum since the late nineteenth century, and by the research stations of the national Meteorological and Hydrological Service (founded in 1911, but closed in 1991), the PAS Institute of Geography (1948), the PAS Nature Protection Centre (1951), and the National Park (1980).

In 1962, a major monograph, synthesizing all research to date, was published (Szafer, 1962). An updated and greatly expanded synthesis has been prepared, but has not been published due to lack of funds. However, partial syntheses are provided by the special issue of *Parki Narodowe i Rezerwaty Przyrody* (Vol. 5, No. 1, 1984) for the National Park's 30th anniversary; the Atlas of the National Park, with thematic maps at scales of 1:50,000 and larger (Trafas, 1985); and a special issue of *Mountain Research and Development* (Kotarba, 1992). All of these contain papers

Figure 11.5 Ski-lift and hiking trail on the west side of Kasprowy Wierch, Tatra National Park, Poland

deriving primarily from research in a range of natural science disciplines, with a few papers on nature conservation and the impacts of various human activities.

One measure of the great depth and volume of scientific research is that 300–400 scientists, working on about 100 projects, have applied each year for permits to the director of the scientific station since the early 1980s. Together with the register of permits kept since 1957, the short reports required for the renewal of permits provide a valuable record of research. In recent years, research has been conducted in probably every natural science field discipline, and no attempt will be made to summarize the breadth of this work. Perhaps the greatest depth of detailed research in recent years has been in the fields of botany, geology and geomorphology (e.g. Figures 11.6 and 11.7). Many projects are very long-term; for instance, some current botanical and entomological studies began in the mid-1950s. The majority of projects are basic natural science research, conducted by scientists for the PAS and universities, in roughly equal numbers, with little involvement from National Park scientists.

The smaller number of applied projects directly involve National Park scientists, who undertake both short- and long-term monitoring studies and also initiate projects (often undertaken by university students) and encourage projects of direct relevance for park management, often at the suggestion of the National Park's Scientific Advisory Committee. These

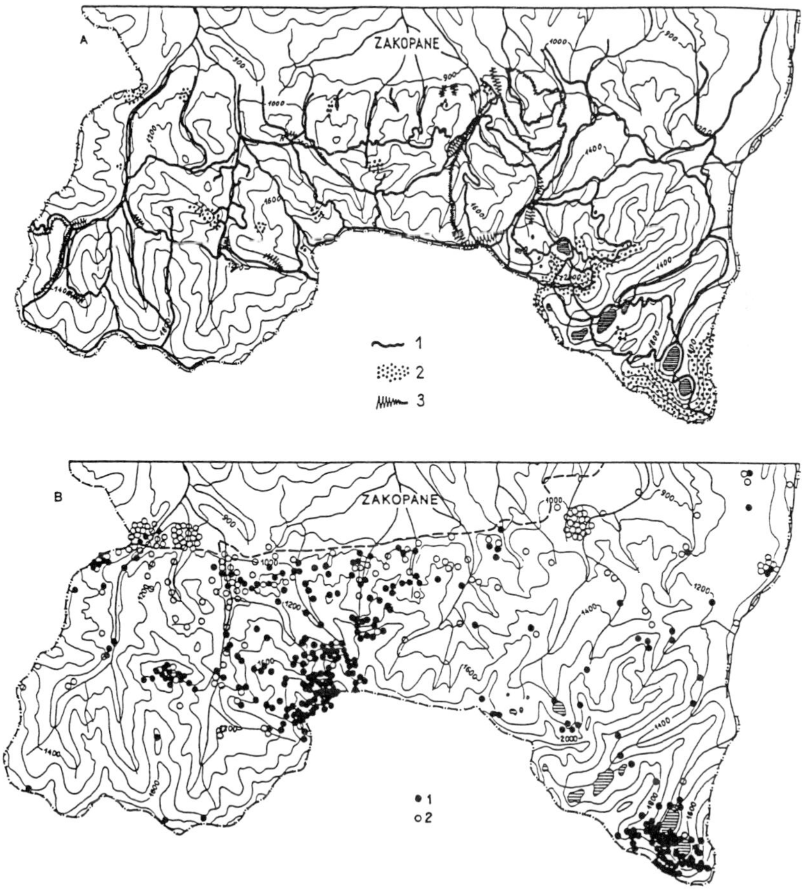

Figure 11.6 Conflict between conservation and development in Tatra National Park. A: areas of tourist pressure; 1 = tourist routes, 2 = mountain climbing areas, 3 = ski areas. B: localities of rare and relict taxa: 1 = mountain species, 2 = lowland and lowland–mountain species (from Mirek, 1992, with permission)

include studies of the impacts of human activities – including sheep-grazing, skiing and tourism – on various ecosystem components. Among the ongoing monitoring studies are two on levels of air pollution and its impacts on coniferous trees at thirty-six sites, which began in 1985 and are now undertaken in conjunction with similar studies conducted by the scientists of the Slovak Tatra National Park. The amount of trans-border research and monitoring has been increasing since restrictions on travel were lifted in 1989 (they were imposed in 1981). In addition, a few scientists from other European countries work in the National Park each year. Most of these

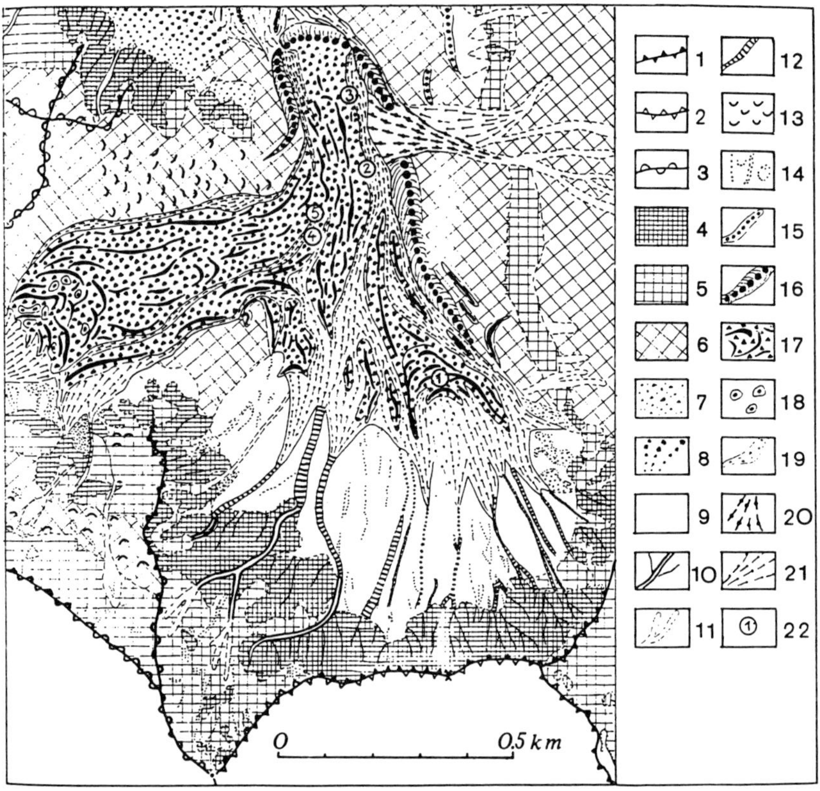

Figure 11.7 Morphology of the Starorobocianski cirque, western Tatras (from Kotarba, 1992, with permission) 1 – narrow rocky ridge crest; 2 – narrow rounded ridge crest; 3 – wide and rounded ridge crest covered with debris and vegetation; 4 – rocky slope and rockfall; 5 – debris-mantled slope; 6 – slope covered with glacial and solifluction deposits; 7 – fresh rockfall talus; 8 – large and rapid rockfall; 9 – rockfall and alluvial slopes of Holocene age; 10 – rocky chute and gorge; 11 – slope trough and gully; 12 – debris flow gully; 13 – solifluction lobe (geliflucion); 14 – tongue, ridge and debris stream of cryonival and pluvio-gravitational origin; 15 – moraine ridge; 16 – moraine ridge related to last recessional phase; 17 – relict rock glacier, partly protalus rampart; 18 – dead-ice hollow; 19 – depression and trough filled with debris flow material; 20 – debris flow cone of late glacial age; 21 – debris flow cone of Holocene age; and 22 – location of geological outcrops

are basic studies, although some consider applied themes, such as the impacts of acid precipitation.

CONCLUSIONS

In spite of the lack of formal MAB-6 activities in Poland, a vast amount of research has been undertaken in mountain areas during the period of the

MAB programme. This includes not only the research summarized above, but also studies in smaller mountain areas such as the Gorce and the Pieniny, for which a major monograph was published prior to the construction of a major dam (Zarzycki, 1982). However, most research in Poland's mountain areas has been conducted within narrow disciplinary boundaries and predominantly in the natural sciences; MAB does not appear to have had any significant influence until almost the present time.

Apart from the usual problems of developing flexible interdisciplinary research programmes in socialist countries, there appear to have been two specific problems in Poland. The first was that the majority of members of the Academy of Sciences did not appreciate, and did not co-operate with, the SocMAB framework for research, imposed from the Soviet Union. The second is that the individual responsible for MAB-6 until the mid-1980s was, from 1979, not within one of the country's major research institutions concerned with mountain areas – although from 1969 to 1979 he was director of an institution whose members have undertaken much research in mountain areas and still do so, even in the current difficult economic period.

Today, in spite of decreasing budgets for scientific research – for instance, the total budget of the Academy of Sciences for 1992 was about half of that for 1991 – the future of MAB activities in Poland's mountain areas looks far healthier than before. The reconstitution of a subcommittee for MAB-6, explicitly linked to research in mountain biosphere reserves and with the added catalyst of organizing a major conference in 1993, means that substantial attention is being given to developing a network of mountain scientists from diverse institutions. This will be linked to the international community through joint activities in the new biosphere reserves and, presumably, as a result of the Zakopane conference. The economic situation also appears to have some beneficial effects, such as greater interaction between park management authorities and scientists who are interested in undertaking applied research. Finally, Polish scientists have been successful in attracting international funding for research and monitoring projects in mountain areas; a necessity in the current period. It is to be hoped that the designation of three new biosphere reserves and the Zakopane conference in 1993 are indicators of the increasing strength of MAB in the mountains of Poland.

CHAPTER 12

BULGARIA

MOUNTAIN REGIONS

Bulgaria's mountains can be divided into two main regions: the Stara Planina (Balkan range) and the south-western mountains (Figure 12.1). The Stara Planina cross the country from the western, Serbian border, dividing the Danubian Plain to the north from the southern half of the country, whose rivers drain south into Greece and Turkey. The highest peak of the Stara Planina is Botev (2,376 m); only the highest altitudes are occupied by

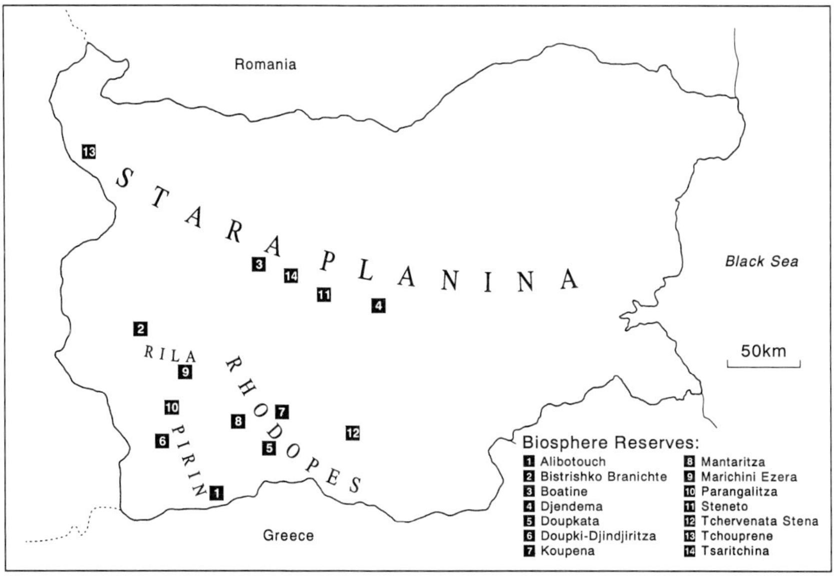

Figure 12.1 Bulgaria: mountain ranges and biosphere reserves

subalpine meadows, whose extent has been increased by centuries of grazing. The other mountains occupy the south-western corner of the country, on the borders with Macedonia and Greece. Along the Macedonian border are the western border mountains, culminating in the Belasitza massif (highest peak 2,029 m) at their southern end, on the Greek border. The other mountains are on the east side of the Struma River. Furthest north is the outlying Mount Vitosha (2,277 m), south of Sofia. Next to the south are the Rila (highest peak: Musala, 2,925 m) and the Pirin (Vihren, 2,915 m). All of these mountains have extensive areas above timberline (*c.* 2,000 m). The largest of the south-western ranges is the Rhodopes (Perelik, 2,191 m), east of the Rila and Pirin mountains along the Greek border.

The current issues of importance in the two mountain regions differ considerably. The Stara Planina is characterized by declining and ageing populations, widespread but decreasing grazing of sheep and goats, a declining forestry industry and little tourism. The south-western mountains can be divided into two groups. The Rila and Pirin massifs are largely protected as national parks, and have low densities of settlement. Both agriculture and forestry are limited. Tourism, in both winter and summer, is an increasingly important component of the economy. In contrast, populations of both people and domesticated animals in the Rhodopes are static or even increasing, the forestry industry is stable, and tourism is growing, particularly in the western, higher parts. A good synthesis of information about all mountain regions was prepared in the 1980s by members of the Institute of Geography of the Bulgarian Academy of Sciences (BAS) and other scientists, particularly from the Department of Geography at Sofia University. The first volume covers topography, climate, water resources, soils, flora and fauna, landscape types and recreational potential (Mishev et al., 1989). The second volume, discussing human geographical and economic themes, has not been published.

ORGANIZATION OF MAB ACTIVITIES IN MOUNTAIN AREAS

The National MAB Committee was established within the national Commission for UNESCO in 1975. Responsibility for the direction of MAB was given to the National Co-ordination Centre for Ecology and Environmental Problems (NCCEEP), which was under the jurisdiction of the Committee for Environmental Protection of the Council of Ministers. The Director of the Centre became the chairman of the national MAB committee. Subcommittees for different MAB Projects were not established. In 1989, after the change of political system, the NCCEEP was abolished and its activities transferred to a new Institute of Ecology in the BAS, which became responsible for MAB. With the exception of short-term projects specifically requested by government departments, all financial support for MAB activities has been from the budgets of participating institutes of the BAS and the former NCCEEP.

A notable feature of the MAB programme in the mountains of Bulgaria is the large number of biosphere reserves. These were chosen to encompass as much of the great diversity of Bulgaria's mountain ecosystems as possible, and all were initially established in 1977 (Nedialkov, 1986). Their legal and administrative status is somewhat complicated. While they are under the jurisdiction of the Ministry of Environment – which established them and makes decisions about enlarging them – management and personnel are the responsibility of the Committee of Forests of the Council of Ministers. There are fourteen mountain biosphere reserves (see Figure 12.1):

(1) Two in the Pirin mountains (Alibotouch, Doupki-Djindiritza);

(2) Four in the Rhodopes (Tchervenata Stena, Doupkata, Koupena, Mantaritza);

(3) Two in the Rila mountains (Marichini Ezera, Parangalitza);

(4) Five in the Stara Planina (Boatine, Tchouprene, Djendema, Steneto, Tsaritchina); and

(5) Bistrishko Branichte on Mount Vitosha.

All of the biosphere reserves are quite small: some are less than 1,000 ha and none is larger than 3,500 ha. However, many have been enlarged in the 1990s, and further increases in size are planned. All are managed as strict nature reserves, sometimes with a buffer zone in which all activities, e.g. grazing, forestry, recreation, are strongly controlled. Also, many are inside the Pirin, Rila and the Central Balkans National Parks. The latter, including all of the Stara Planina reserves except Tchouprene, was established in 1991, and has an area of 73,000 ha. However, a park administration is not yet in place; the park continues to be managed by the foresters responsible for each of its districts. The Rila National Park was also extended to more than 100,000 ha in 1992; the Pirin National Park is just over 40,000 ha in size. As all MAB research in Bulgaria's mountains has taken place in biosphere reserves, and all but three are in mountain areas – one of the latter is in the low Strandja mountains, near the Black Sea coast – Projects 6 and 8 have generally been treated as one.

SCIENTIFIC RESEARCH IN MOUNTAIN BIOSPHERE RESERVES

In the early 1980s, four meetings brought together scientists working in mountain biosphere reserves and other protected areas in Bulgaria. The first, in 1981 and 1982, were national conferences on biosphere reserves and nature protection. Unfortunately, proceedings of these meetings could not be located. In October 1983, an international conference on high-mountain ecosystems was held in Vratsa. The proceedings were published in three volumes (Bulgarian National Committee, 1983), with the following sections:

(1) General ecological problems of mountain ecosystems;

(2) Ecological–economic and settlement problems of mountain ecosystems;

(3) Mountain forests and recreational ecosystems;

(4) Mountain agricultural and meadow ecosystems; and

(5) Mountain ecosystems and ecological education.

In September 1985, a conference was held in Blagoevgrad, on the conservation of natural areas and the genetic material they contain, resulting in three volumes of proceedings (Nedialkov et al., 1985). Analysis of the English summaries of the papers in the two sets of proceedings suggests that most were based on short-term studies – not always in mountain areas – and that these were conducted by independent scientists who were not part of interdisciplinary research programmes. As discussed below, multidisciplinary studies, predominantly by natural scientists, only appear to have been undertaken in a few mountain biosphere reserves.

Other than the two sets of proceedings, there are few published sources of information about scientific research in Bulgaria's mountain biosphere reserves. Much of the work conducted by members of the BAS' Institute of Forestry Research has been published in their quarterly journal *Nauka Zagorata* (Forest Science) and in the proceedings of forestry conferences (e.g. Tsankov et al., 1984; Velkov et al., 1988). Some publications are at various stages of preparation, a process which has always been limited by available funds. Still other studies have been prepared as internal reports for different ministries and government committees, and there are no plans for their publication.

The lack of published scientific findings means that the following synthesis of research in mountain biosphere reserves is drawn largely from extensive discussions with scientists in many institutes and academic and government departments (see Author's Acknowledgements), together with analysis of the two sets of conference proceedings. However, given the general lack of communication and collaboration between, and even within, scientific and academic institutions, and the limited time available for this review, it is likely that the information below is not quite complete. Where institutes are referred to, these are institutes of the BAS.

Pirin

The flora of the Pirin mountains includes over 500 vascular species, many of them endemic. In addition, the forests are extremely productive, and include the oldest trees in Bulgaria. Scientists from the Institute of Forestry Research have undertaken research on tree growth, nutrient cycling, succession, pathogens and insect pests in the Doupki-Djirinditza reserve. This reserve is also the location of ongoing studies of birds (Institute of Zoology) and flora and vegetation dynamics (Institute of Botany). In the Alibotouch reserve, research has been limited to studies on the structure and dynamics of the pine forests (Institute of Forestry Research) and on floristics (Institute of Botany).

Rhodopes

Of the four reserves in the Rhodopes, only Doupkata has been the subject of much scientific research, conducted by the Institute of Forestry Research, and considering forest structure and function. Scientists from the institute are currently undertaking a short-term programme, at the request of the Committee on Forests of the Council of Ministers, on forest genetics, entomology and pathogens. The aim of the study is to assess the biophysical status of the Scots pine and spruce forests; considerable numbers of trees are showing signs of decline. In Tchervenata Stena, ornithological research has continued since the mid-1980s. Floristic surveys have been conducted in all reserves in the same period by scientists from the Institute of Botany.

In 1992, the National Fund for Scientific Investigation provided support to the Institute of Ecology and other institutes, particularly the Institute of Urbanism and Architecture, for a 3-year project on conflicts between nature protection and local economic development, in areas being developed for tourism in the western Rhodopes.

Rila

The two oldest research stations of the Institute of Forestry Research were established in the Rila mountains in 1961 and 1963. The spruce forests of Parangalitza reserve are the most productive in Europe. Since 1978, about twenty scientists from the Institute of Forestry Research and eight other institutes have conducted research in these forests. The main goal of this research is to establish the structure and functioning of these forests. Studies have focused on nutrient uptake and cycling, competition, succession, tree growth, pathogens, insect pests, soil microbiology, meteorology and hydrology. The results have been published in the proceedings of the 1983 Vratsa and 1985 Blagoevgrad conferences and those of other international conferences (e.g. Tsankov et al., 1984; Velkov et al., 1988), and also in a book on the water balance of spruce forests (Prax and Raev, 1985). This work continues, and has recently been given impetus by special funds obtained in 1991 from the BAS for a 2-year in-depth project, involving thirty-five scientists from various institutes, comparing Parangalitsa with Bistrishko Branichte, Steneto and Strandja reserves.

Less research has been done in the other reserve, Marichini Ezera, in spite of its floristic richness; a list of over 500 vascular plant species from both coniferous forests and the extensive subalpine and alpine zones has been compiled by scientists from the Institute of Botany, who have also conducted research on vegetation dynamics and succession.

In 1992, the Institute of Ecology received support from the National Fund for Scientific Investigation to begin a 3-year project on conflicts between nature protection and local economic development in areas being developed for tourism in the Rila mountains. This project also involves other institutes, particularly the Institute of Urbanism and Architecture.

Stara Planina

A considerable amount of research has been done in the reserves of the Stara Planina, particularly in connection with the development of the new Central Balkans National Park. The Institute of Forestry Research established a field station near Steneto in the early 1970s (Figure 12.2), with two permanent staff (only one since 1991).

Most of the forests of the Boatine reserve are beech, with spruce at higher elevations. Research by scientists for the Institute of Forestry Research has concentrated on pedology, microclimatology, tree metabolism, organic matter uptake and forest succession following canopy opening by severe winds. Scientists from the Institute of Ecology have undertaken work on forest composition and changes in vegetation composition resulting from grazing, while studies of the distribution of birds and large mammals have been conducted by scientists from both the Institute of Ecology and the Polish Academy of Sciences (PAS), beginning in 1978. A synthesis report based on all research in the reserve will soon be published by the Nature Conservation Research Centre of the PAS in Krakow.

The Tsaritchina reserve has the widest range of ecosystems of any in the Stara Planina, from beech forests at the lowest elevations to subalpine meadows (Figure 12.3). Most research in this reserve has been conducted by scientists from the Institute of Ecology and the PAS Nature Conservation Research Centre in Krakow, who began work in the reserve in 1981. In

Figure 12.2 Information signs at the boundary of Steneto biosphere reserve

Figure 12.3 Upper pine and fir forests and subalpine grassland in Tsaritchina biosphere reserve

addition, scientists from the Institute of Forestry Research have been conducting research on forest productivity and structure, and on the influence of vegetation on soil evolution, since the mid-1970s. The initial results of this work, which continues, have been compiled in a report, published in Krakow, with chapters on soils, geology, geomorphology, mammals, birds, flora, ecosystem structure and tree biomass (Klimek, 1987).

Less scientific research has been undertaken in the other three reserves. Research on the structure and processes of the forests of the Steneto reserve has been conducted by scientists of the Institute of Forestry Research, the Slovak Academy of Sciences and the Institute for Forest Ecology in Zvolen, Slovakia. In 1991, work began on the structure and function, and particularly water balance, of spruce forests, as part of a 2-year project comparing this reserve with Bistrishko Branichte, Parangalitza and Strandja. Baseline research and monitoring of soils, tree populations and hydrology is conducted by scientists from the Institute of Ecology. Scientists from this institute also conduct ongoing research on bird and large mammal populations. Similar work is conducted in Djendema reserve. In the Tchouprene reserve, scientists of the Institute of Forestry Research have conducted some research on the genetics of the predominant spruce forests, and surveys of bird species have been undertaken by scientists from the Institute of Zoology.

Vitosha

Scientists from the Institute of Forestry Research have conducted research in the spruce forests of Bistrishko Branichte reserve since the late 1970s. This ongoing research considers the genetics of tree populations, organic matter accumulation and microflora in soils, meteorology and groundwater hydrology. Scientists from the Institute of Botany have conducted long-term studies of changes in vegetation resulting from recreational use, based on research dating back to the first decades of the century. In 1991, work began on the structure and function, and particularly water balance, of spruce forests, as part of a 2-year project comparing this reserve with Parangalitza, Steneto and Strandja.

CONCLUSIONS

The assessment and synthesis of scientific research in Bulgaria's mountain biosphere reserves is not easy and, as Bulgarian scientists recognize, is complicated by the inaccessibility of their language. Consequently, the recent and forthcoming syntheses for Boatine and Tsaritchina, published in Krakow, are in Russian with English summaries. Similar syntheses should be prepared for other biosphere reserves. Another critical difficulty in reviewing MAB activities in mountain areas is that they have been undertaken by scientists from a large number of institutes, with little collaboration. Throughout the period of the MAB programme, until the present, there appears to have been little communication between scientists from different institutes; and, in some cases, even from the same institute. This 'balkanization' of science is paralleled in government departments, most obviously with regard to the administration and management of biosphere reserves by both the Ministry of Environment and the Committee on Forests of the Council of Ministers, each of which tends to work mainly with a specific set of institutes. Another result of this bureaucratic complexity is that research in a biosphere reserve requires permits from both of these agencies and the Academy of Sciences.

Overall, neither the zonation and management of the biosphere reserves nor the research done within them can be said to have shown great adherence to the concepts of the MAB programme. The reserves consist of a core zone (and sometimes a buffer zone), and the research has rarely been multidisciplinary, let alone interdisciplinary, and has been conducted exclusively by natural scientists. The new projects being undertaken by scientists in the Institute of Ecology in the Rila and Rhodopes are a positive sign. Scientists from this institute have recently prepared proposals for research, involving both natural and social scientists, on the development of mountain areas in the transition to a market economy, but these are still on the drawing board. In addition, the new and expanded national parks provide opportunities for applied research comparing both natural and managed ecosystems.

In conclusion, it is difficult to say that the MAB programme has had much significance in the mountains of Bulgaria, with respect to either research or resource management. All of the biosphere reserves were already protected before their inclusion in the MAB network. No specific funds were ever provided for MAB activities, Bulgarian scientists appear to have had little exposure to the concepts of MAB, and 'balkanization' – rather than multi- or interdisciplinarity – has been the rule. While shrinking budgets for science can be a stimulus to improving communication and co-operation among institutes and government agencies, such interactions are scarcely apparent in Bulgaria, though they are needed in a country whose mountains are a vitally important environmental and economic resource.

PART III

MAB from the Pyrenees to Siberia: an evaluation

CHAPTER 13

MAB FROM THE PYRENEES TO SIBERIA: AN EVALUATION

The range of the research activities, conducted over more than two decades and presented in the previous chapters, may be compared to the great environmental and cultural diversity of the mountain regions in which they were undertaken. This concluding chapter compares these very diverse activities and draws some common lessons from them.

The main point of comparison relates to the proposals of the MAB-6 working group in Lillehammer in 1973 (UNESCO, 1974). To recapitulate, the general objective of MAB-6 activities in temperate mountains was to gain a better knowledge of these ecosystems in relation to human use, in order to conserve present ecosystems and devise new stable ecosystems to replace old systems that are no longer socially relevant and economically viable. Research was to be undertaken within spatially-integrated units, giving attention to the temporal dimension.

A proposed focus for research, to be undertaken co-operatively by social and natural scientists, was the impacts of tourism. This would require the general description of:

(1) Ecological systems;

(2) Patterns of impact; and

(3) Reciprocal relationships between human beings and their environments.

Two further issues were to be emphasized, critical factors and threshold points in the process of change and the needs of local people and their control of the environment, or the means by which this control is lost.

Twenty years later, this concluding chapter evaluates the extent to which these objectives were realized in the design and implementation of MAB mountain activities in Europe and the former USSR. The chapter has four sections. First, it traces the evolution of models in the MAB-6 programme. Second, it explores organizational and institutional issues relating to the course of different projects and activities. Third, it presents some conclusions which may be of use in the design and implementation of future interdisciplinary research programmes. It concludes with a consideration of the legacy of MAB-6 in the mountains of Europe and the former USSR.

MODELS IN MAB-6 PROJECTS

The Lillehammer group proposed that the basic objectives of the MAB-6 programme in temperate mountains should be fulfilled through the use of a model which identified 'the holistic inter-relationships of man and his mountain ecosystem' (UNESCO, 1974: 34). It was hoped that such a model could be used and replicated in comparative international research. This section traces the evolution of models, beginning with those presented at the Lillehammer workshop, and draws on the reviews of mountain models presented by Ives and Messerli (1990) and, for the Swiss MAB-6 programme, by Messerli (1986).

The discussion follows a sequence from the Austrian Obergurgl project to the Swiss MAB-6 programme and then to the projects in the Catalan Pyrenees (Spain) and Berchtesgaden (Germany). Other conceptual models have been developed during European MAB-6 projects, notably in France (Briançonnais and Inter-parks) and Spain (Aragon Pyrenees). These did not attempt to be holistic, focusing on the pastoral agricultural and forestry components of regional human–environment systems, with a strong concentration on their natural scientific aspects. However, they have been valuable in developing management regimes. The process-oriented studies in the Grossglockner and Gastein areas of Austria also resulted in models, but these are strongly ecological, including human beings merely as agents of change, if at all.

Obergurgl (Austria)

As noted in the chapter on Austria, the conceptual model used in the Obergurgl project derived from a systems-analytical approach developed during previous activities within the International Biological Programme (IBP) in the area, together with the project co-ordinator's lifelong experience in the area. Two simple precursors to the Obergurgl model were presented by the project co-ordinator at the Lillehammer workshop, one dealing with the internal system of tourism in a mountain valley, the other with external influences (Figures 13.1 and 13.2).

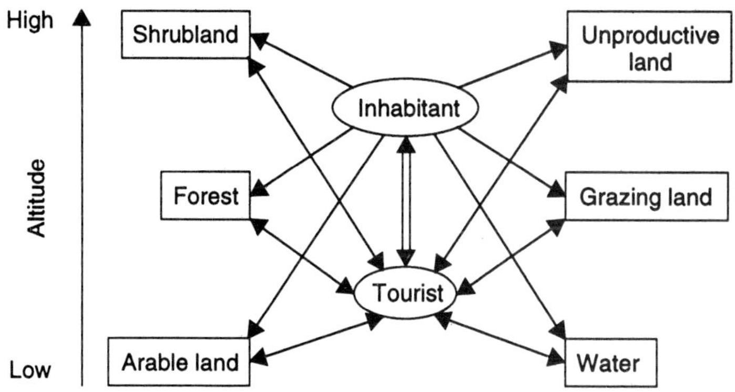

Figure 13.1 Schematic framework for tourism research: internal sector (from UNESCO, 1974, with permission)

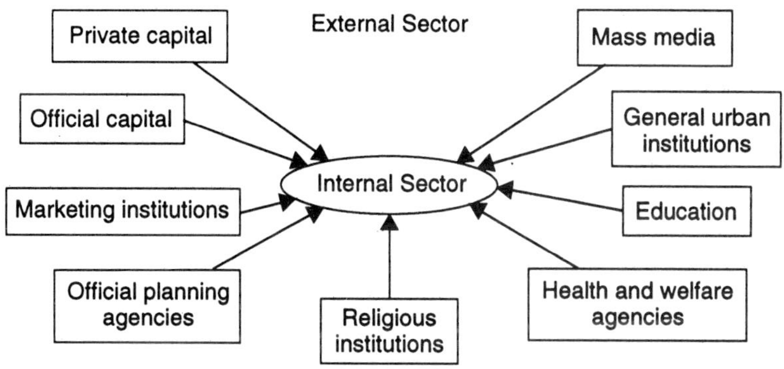

Figure 13.2 Main elements of external sector impinging on internal sector (from UNESCO, 1974, with permission)

These simple models were combined and expanded at the 1974 workshop at the International Institute for Applied Systems Analysis (IIASA), in which scientists, government officials and people from Obergurgl took part. The workshop had three objectives (Himamowa, 1974: 3):

(1) To promote communication among the interest groups involved in the Obergurgl project, by using the simulation model to provide a common language and focus for attention;

(2) To define, through data requirements for the model, critical research areas for the MAB-6 project; and

(3) To provide alternative long-range (20–40 years) forecasts for the people of Obergurgl concerning likely impacts of various development strategies that they considered practical.

The final model (Figure 3.3) was developed by four sub-groups of participants, which considered four general topics: recreational demand; population and economic development; farming and ecological change and land use and development control.

This modelling exercise used a variety of quantitative data, together with simple assumptions derived from scientific research and local knowledge. The ranked priorities for further research revealed many of the gaps in this initial set of data and assumptions. Five of these priorities, including the two highest-ranked, required research by social scientists with regard to: the attitudes of villagers and tourists (1 and 2); the projection of potential recreational demands (5); evaluation of development alternatives and research priorities (6); and economic analysis (8). However, the lack of many necessary natural scientific data was also recognized in the three remaining priorities: ecological mapping (3); determination of primary production in relation to grazing (4); and experimental ecological studies relating to grazing, trampling and construction (7).

In 1975, the original model was reduced to its most important components, in order to simplify the integration of new data deriving from the research activities and to provide a better basis for discussion with local people (Figure 3.4) (Moser, 1987). However, while the early research in the MAB-6 project was mainly conducted by natural scientists, most of the components of the simplified model required sociological and socio-economic research. These components varied from very complex topics, such as landscape quality and carrying capacity, to relatively simple demographic and economic measures. The relationship of the simplified model to the research then underway is unclear; this model appears to have been something of a dead end. The later revision in 1978, might have remedied these problems; but no published outline of this version exists, and the project was terminated the year after.

In spite of the premature demise of the Obergurgl project, the interactive development of the model by scientists and local people had a number of positive results both in Obergurgl and elsewhere. Though many of the findings of the research were preliminary, many environmental and economic constraints on the continued development of tourism were identified; the need to concentrate on the quality of tourism, rather than the quantity of tourists, was recognized; and the stresses on local families and institutions were revealed. These findings led to various decisions, at least in the short term: for instance, to limit accommodation and ski-lift capacity, minimize the impacts of hiking, revegetate ski-runs, provide free transport in winter, and establish cultural institutions and activities (Moser, 1987). Within Austria, innovative land-use planning also derived from the project; and the initial Obergurgl model provided a basis for MAB-6 activities in other countries, as discussed below.

Swiss MAB-6 programme

In the later stages of the Obergurgl project, there were many interactions between participating scientists and those involved in developing the Swiss MAB-6 programme, at meetings in Austria, Germany and Switzerland. In addition, the co-ordinator of the Swiss programme spent some time at institutions in Germany which had developed quantitative sensitivity models to investigate the future of ecological–social systems. Drawing on the experiences and the results of the 1977 feasibility study in Grindelwald, a new simplified model was developed in 1978, incorporating both internally- and externally-driven processes (Figure 4.2) (Messerli and Messerli, 1978). The authors stressed that this was to be regarded as a first attempt to clarify which relationships should be investigated during the MAB-6 programme. In modelling terms, the variable to be optimized was the land-use subsystem: the link between the constraints of the natural system and the demands of the socio-economic system.

Based on this simplified model, the groups working in the different test areas chose a variety of modelling approaches, with respect to both the storage of spatial data and the treatment of quantitative data. Two approaches were chosen for storing spatial data, using some of the first geographic information systems (GIS) developed for mountain areas: homogeneous polygons in Grindelwald and grid-cells in Davos. In Grindelwald, simulation modelling was conducted using both input–output and systems–dynamic models, while linear-programming simulation and dynamic balance models were used in the Pays d'Enhaut. Another group of models should also be mentioned: models of the incorporation of local people in the definition of research and the evaluation of results. In only one project, in the Pays d'Enhaut, were local people strongly integrated from the beginning. There were also connections to the local authorities in the other areas, but involvement of the local population in general was not emphasized, although publicity was provided in newspapers, and meetings were held as results appeared.

Conclusions about the success of the modelling approaches used in the different MAB-6 projects were presented by Messerli (1986) and are briefly summarized below, bearing in mind his comment that 'there were few simple indicators for assessing whether we succeeded or failed in our objectives and, as the process evolved, our original expectations adjusted to the established reality' (p. 146). This was particularly true because the topics and methodologies of research and models varied so greatly between the test areas. An evaluation of overall 'success' might have been easier had standardized data sets been collected for all areas, but this idea was strongly resisted by the project co-ordinators. Yet, the range of approaches did allow a comparative evaluation that was, perhaps, more valuable for the long-term methodological development of modelling in complex human-environment systems.

Messerli (1986: 153–6) presented five conclusions regarding the models used in the Swiss MAB-6 programme:

(1) The overall model was partially operationalized, showing the effects of various hypotheses of socio-economic development on land use and thus the natural environment. The extent of different economic, socio-cultural, and ecological changes could be qualitatively estimated. Significant linkages were shown in the socio-economic realm by simulation modelling in Grindelwald and the Pays d'Enhaut, and between land-use and natural systems in the Davos landscape model. However, it was not possible to establish relationships between environmental and recreational quality and the behaviour of residents and tourists, although supposed patterns of behaviour were elucidated by studies of attitudes and aesthetic perceptions.

(2) The socio-cultural system was shown to be central in defining human-environment relationships. While economic factors were important in spatially-orientated modes of behaviour and decisions relating to land use, these were strongly influenced by the cultural models of regional actors.

(3) The choice and structure of spatial reference systems for linking socio-economic and ecological data greatly affected possibilities for interdisciplinary work. A flexible, visual GIS was very valuable for communication and research.

(4) The simulation models produced indispensable quantitative information. They stimulated research by provoking new questions, uncovering data gaps and posing problems of linkage and quantification. After construction, they supplied illustrative data for developing and demonstrating different scenarios. A modular structure allowed adaptation to new scenario conditions. However, only economic and demographic relationships could be quantified.

(5) In spite of the different methodological approaches to simulation modelling ('top-down' in Grindelwald, 'bottom-up' in the Pays d'Enhaut), comparable findings were produced with regard to the sensitivity of regional human-environment systems to development scenarios.

A further conclusion with regard to the models of the Swiss MAB-6 programme relates to the interaction of local people with scientists in the test areas, particularly bearing in mind that the Lillehammer group stressed that MAB-6 projects should consider the needs of local people and their control of the environment. The Pays d'Enhaut project built on over a decade of scientific involvement in regional planning, and established a unique association to ensure continued interaction. In contrast, the Grindelwald and Davos projects took place with little interaction until research results revealed problems of concern to local people. In Davos, awareness of environmental issues was growing as a result of the efforts of the local forester, and the recognition of significant air pollution heightened this. In Grindelwald, interest developed from socio-cultural research, which brought problems out in the open and led to the establishment of a planning group and many meetings at which simulation models were used to help develop planning guidelines. As the results of the projects in all four test areas were eventually used by local people for planning, one might conclude that their initial lack of involvement was not critical for effective applied research. However, the course of subsequent MAB-6 projects suggests that such involvement is preferable, as discussed below.

Further applications of Swiss MAB-6 models

Knowledge of the models developed in the Swiss MAB-6 programme has been widely disseminated. Two contrasting applications, in the Catalan Pyrenees and Berchtesgaden, are discussed below. Drawing substantially from the experience of the Pays d'Enhaut project, the Catalan project was co-ordinated by local people and scientists (either born or based in the area), the research concentrated on developing an understanding of the local socio-economic-cultural system, and the local population was kept well-informed. As in the Pays d'Enhaut, while there were comparatively few natural scientific studies, focusing mainly on changes in land use and forestry, the trends identified from these historical changes were valuable in developing scenarios of possible futures for local communities. The project was explicitly intended to provide information relevant for regional planning, and many recommended actions have taken place although, as in the Pays d'Enhaut, it is not possible to definitively attribute these to the MAB-6 project. The project co-ordinators hoped that the data collected during the project would be used for simulation modelling, using linear programming, but resources unfortunately did not permit this.

In contrast, the Berchtesgaden project drew from the conceptual and computer modelling approaches developed in Switzerland, but local people were little involved except for the special studies on the Winter Olympics and agriculture. In addition, the project co-ordination team was strongly weighted towards an understanding of processes within the framework of landscape ecology. The team did not greatly emphasize changes over time or, particularly, the resource management or planning aspects of the work. This is despite the fact that nearly half of the test area is a National Park, while the rest has a tourist-based economy with continuing agriculture and forestry.

One of the principal conclusions of the Swiss MAB-6 programme was that a regional human-environment system cannot be considered only in terms of the internal sub-systems and external factors included in the initial model. Just as important, different sets of processes act at various levels, each requiring specific types of data and research methods. Thus, the 'external factors' of the original Swiss model were better portrayed as affecting each internal sub-system at different levels. These conclusions were operationalized in the Berchtesgaden model (Figures 5.4 and 5.5), which was initially developed in the later stages of the Swiss MAB-6 programme through frequent interaction between scientists participating in the two programmes.

The hierarchical systems approach taken in the Berchtesgaden project retains the internal sub-systems defined in the Swiss model but treats them at three different levels above the 'reality' level of energy flows and material cycles and interactions. It is worth emphasizing that the Berchtesgaden modelling approaches not only represented conceptual evolution from those developed during the Swiss project, but were made possible by significant advances in computer technology.

The three levels of the Berchtesgaden model are as follows:

(1) The 'process' level, at which 'reality' is characterized. The results (usually quantitative) of research obtained at this level are stored in a GIS based on polygons with homogeneous properties, permitting calculations of inputs and outputs of energy and materials to be calculated with an ecological balance model.

(2) The 'dynamic' level, in which dynamic models using aggregated data – some quantitative, some qualitative – are used to predict temporal trends for different variables. These models included a regional feedback model and others which evaluated trends in forest damage and the impacts of the Winter Olympic Games.

(3) The 'strategic' level, which is strongly influenced by external events which may be highly uncertain. Scenarios developed at this level are generally used to drive simulations of resulting impacts at lower levels.

Linkages between the three levels of the model were made by a 'soft coupling' process. This 'means that an expert or group of experts with good local knowledge of the study area advise on how the data collection and the model development must be related together. They also ensure that the overall model, sectors and modules do not become too complex. They interpret data, or aspects of model structure, which may in some situations be numerically similar or have different meanings' (Ashdown et al., 1988).

In summary, the Berchtesgaden project went far beyond its predecessors in the application of computer technology and the development of conceptual models: though, being fixed quite early on, these did not keep up with recent advances in ecological theory. The massive volume of data collected and stored in the state-of-the-art GIS has, to date, principally been used to describe, simulate and monitor trends in selected components of the regional human-environment system, especially in the National Park; much less in the 'forefield' which comprises the rest of the test area (and biosphere reserve). Apart from the development of the GIS, perhaps the most successful element of the project was the special study on the Winter Olympics. This involved many local people over a 2-year period, and came to a number of conclusions regarding the future development of the region. However, it focused primarily on the economic, demographic, transport and tourist sectors. Thus, while computer-based models of both landscapes and socio-economic systems were successfully developed, these two types of models do not appear to have been operationally linked.

Consequently, it is perhaps worth asking a rhetorical question: how might the project have developed, had the group which set its objectives included members of local communities? In the early years of the project, relationships between the National Park authorities and local people were not close, though they have improved in recent years. Consequently, the biosphere reserve remains, in many ways, more conceptual than real. The real test of the linked group of Berchtesgaden models is still to come: their use in developing scenarios and plans for the entire reserve.

To conclude, the modelling approaches developed in the MAB-6 programme, from Obergurgl to Berchtesgaden, have shown great conceptual evolution, aided particularly by advances in computer technology. However, the useful application of this greater scientific knowledge remains limited by the need to integrate effective models of local participation in the design and evaluation of interdisciplinary research projects.

ORGANIZATIONAL AND INSTITUTIONAL ISSUES

Surveying the diverse activities that took place within MAB-6 projects and in mountain biosphere reserves in Europe and the former USSR, one can conclude that the institutional frameworks of these activities strongly influenced their results. While each country had a National MAB Committee, established between 1970 and 1975, this is the only common factor across the countries. This section explores the variation in both the institutional arrangements of MAB-6 programmes at the national level and the internal organization of individual MAB-6 projects.

One primary contrast between the National MAB Committees in the former Communist countries and in western Europe relates to their internal division of responsibilities. While MAB-6 sub-committees were established in each of the former, only one of the latter countries – France – had groups based on MAB Project Areas; and these groups were disbanded during the reorganization of the committee in 1982. However, the Swiss National Committee was never concerned with any topic but MAB-6; and this was also the only focus of the Austrian National Committee in the 1970s.

Central and eastern Europe and the former USSR

Given the existence of MAB-6 sub-committees in the former Communist countries, one might have expected more MAB-6 projects to have taken place. The fact that they did not can be traced to a number of linked factors. First, with the exception of the USSR after 1985 and, to a lesser extent, before 1979, there was generally a lack of leaders (i.e. sub-committee chairmen) who were active in mountain research, preferably based in a strong institution, and respected by their peers. This problem can also be linked to the failure of attempts to develop a co-ordinated research programme at the scale of the socialist bloc.

The other factors are probably a reflection of the political and institutional systems in these countries until the end of the 1980s. Responsibility for MAB activities was generally given to natural scientists; research tended to be formulated and undertaken according to formal – usually 5-year – plans; and there was a tradition of independent research, with little co-operation between institutions – or even between scientists working in different parts of the same institution. Furthermore, the proposed emphasis of the MAB-6 programme on the assessment of small regions, with concern for their future, was at odds with a national centralized system of planning and resource management. Finally, few if any financial resources were specifically provided for MAB activities, from the development of projects through to the publication of reports. Such resources might have stimulated interactions between scientists and institutions, and perhaps have promoted interdisciplinary research involving both social and natural scientists, but it is hard to say.

Thus, while research within MAB-6 projects in the USSR and in mountain biosphere reserves in the other countries was often of a high standard, this was predominantly undertaken by natural scientists. Only in parts of the former USSR (the Caucasus and, to a lesser extent, Kyrgyzstan and Tajikistan), where geographers were involved in the planning and implementation of research, was this somewhat interdisciplinary, working towards the 'official' objectives of MAB-6. It is notable that this research resulted directly from exposure to MAB-6 approaches developed in western

Europe, particularly Switzerland and, later, from direct interactions between 'western' and Soviet scientists during conferences and field-trips in the former USSR. Yet, in general, the lack of such interactions was undoubtedly another hindrance to the evolution of MAB activities in the mountain regions of central and eastern Europe and the former USSR until the 1990s.

Western Europe

In western Europe, there have been significant differences in the organization of MAB-6 projects in the five countries. In France, the country with the greatest diversity of MAB-6 projects, these were 'labelled' but not directly funded by the MAB Committee. These projects have all involved scientists from a number of institutions and different disciplines, and have been financially supported by a wide range of national government agencies, as well as the European Communities. The thematic and disciplinary mix of these projects has varied greatly, reflecting the varying interests of the involved scientists and particularly the research co-ordinators, who have come from a wide range of social and natural science disciplines. In general, French MAB-6 research has become more applied, with greater consideration of societal processes, over time.

Thus, the main value of the French MAB-6 programme has been to catalyse inter-institutional, multidisciplinary projects, as well as to stimulate information exchange and co-ordination at the European scale, through the *Reseau Européen Monde Alpin and ICALPE.* A large number of publications have resulted, though these have been published by agencies sponsoring research, not in a MAB series. Overall, the projects have certainly provided a better understanding of mountain ecosystems in relation to human use, and many critical ecological factors in processes of change have been recognized. However, only in the more recent projects, starting to some extent in the inter-park programme and increasingly in the pre-Alps and the Massif Central, have the needs of local people become a focus of MAB-6 projects.

As in France, support for the initial Spanish MAB-6 projects, in the Aragon Pyrenees, came from usual sources of funding for fundamental research, provided on an annual basis. These projects were 'labelled' as 'MAB', but were largely continuations of ongoing biological research activities, contributing to greater knowledge of the ecological, forest and pastoral systems of the study region. No funds have ever been provided for synthesis or the development of system-level models.

Similarly, the Austrian MAB-6 projects were funded on an annual basis through traditional channels. The funds were limited and, by the late 1970s, there was considerable competition for them. Furthermore, the National

Committee never insisted on interdisciplinary applied research, for which there was no tradition in Austria; with the principal exception of the Grossglockner and Gastein process-orientated studies, the majority of participants in each project came from the co-ordinator's institution. These are all probable reasons for the fact that, with the partial exception of the Obergurgl project, the Austrian MAB-6 projects were neither holistic nor prospective. They essentially provided detailed information about certain components of regional human-environmental systems, with a primary emphasis on ecological systems in the Grossglockner and Gastein projects, and a somewhat greater emphasis on socio-economic systems and their impacts on ecological systems in the Sameralm and Obergurgl projects. While the report series brought the component studies of each project together, there was little attempt at synthesis; overall, the primary applications have been in the regulatory sphere, in the limited domain of downhill skiing.

The remaining MAB-6 projects, in the Catalan Pyrenees, Switzerland and Germany, were all supported with special funds over fixed periods. The Catalan project was perhaps the most innovative of all those discussed in this book, being initially developed outside the official MAB infrastructure and financed by two national ministries interested in policy-orientated research; though the second only provided funding after it found the early results satisfactory. The project was unique in being co-ordinated by both a university scientist and a member of an action-orientated group, with institutional support from the local municipal administration. Greater explicit consideration was given to the needs of local people than in any other MAB-6 project, with a primary emphasis on ensuring the survival of economically and culturally viable communities, supported by a secondary focus on how to manage ecosystems in order to do this. Many recommendations have been implemented. That more have not can be traced, at least partly, to institutional factors. The development of mountain areas is under the jurisdiction of provincial governments, not the national administration (which supported the project and subsequent research in the area). The local authorities do not belong to the party in power in Catalonia. Thus, even when the needs of local people are recognized in applied research projects, the application of their findings often requires political will at a higher level: the 'strategic' level formulated in the Berchtesgaden project.

The two largest national programmes, in Switzerland and Germany, were both preceded by about 4 years of preparatory work. This process began in each country with questionnaire surveys to identify interested scientists and topics for research. In Switzerland, these surveys also helped in the choice of study areas, which was based on: a geographical distribution which ensured a wide range of environments and access to a large number of research institutions; the existence of past and current research; and growing resource management challenges and conflicts. In both countries, the surveys and subsequent activities in the planning stage showed a lack of experience of interdisciplinary research. Since the groups charged with overseeing the projects recognized that this was necessary, appropriate research approaches were redefined in Germany through the preparation of a second feasibility study, and in Switzerland by a field study in one of the proposed test areas.

The Swiss and German projects also varied significantly in their organization, reflecting national differences as well as the fact that there were four test areas in Switzerland and only one in Germany. The Swiss project was actively overseen by an expert group, as required for all national research programmes funded by the Swiss National Foundation. This group was highly multidisciplinary, also including technical practitioners and government representatives, and effectively replaced the National MAB Committee. The overall programme was run by two university co-ordinators, and there was also a senior and a junior co-ordinator for each test area. In contrast, the German MAB Committee provided general oversight for the Berchtesgaden project, but its management was largely left to the project co-ordinating team, including representatives of the National Park Administration, together with other scientists who were alternately based either at a university or in a consulting firm (for reasons specific to Germany, as explained previously).

In Berchtesgaden, the choice of scientists and topics for research was made principally by the co-ordinating team. The requirement to produce data for inclusion in a GIS was a major unifying factor; as it was also in Davos and, to a lesser extent, Grindelwald. In Switzerland, the expert group made decisions regarding participating scientists and their research, and these were then implemented by the programme and project co-ordinators (who were consulted before decisions were made). Thus, with one partial exception, there was a clear hierarchical structure in the organization of the MAB-6 projects in the two countries, with some flexibility maintained by formal and informal meetings of project scientists. The exception was in the Pays d'Enhaut where, after the expert group refused the project team's initial proposal for relative autonomy, they established a research co-operative which maintained strong links to the local community through the MAB-Pays d'Enhaut Association.

Thus, as discussed in the previous section, the integration of research findings into local decision-making occurred at different stages of the MAB-6 projects. In the Pays d'Enhaut, such integration existed from the start of the project. In Grindelwald, it began during the synthesis phase, when the project co-ordinator, together with local administrators and representatives of interest groups, formed a group which prepared planning guidelines which were subsequently approved by 90 per cent of the local population. In Davos, research findings led to the establishment of an air quality monitoring network and, after the project concluded, were used in the development of a master plan for the commune. Similarly, MAB-6 results were recently used for the master plan for the two ski-resorts in the Aletsch area. In Berchtesgaden, results of the MAB project have influenced some local policies, but are only just beginning to be used for the master plan for the National Park.

Conclusions

Reviewing the course and practical application of MAB activities in the mountains of Europe and the former USSR, one can conclude that the ideals of the Lillehammer group, 20 years ago, have been partially realized. Almost all projects in western Europe and the Caucasus provided a better understanding of mountain ecosystems in relation to human use. Yet, while many of these ecological and land-use studies considered present processes and looked backwards, few gave much attention to the future or were integrated with other studies that provided a complementary understanding of the massive cultural, economic and demographic changes that have taken place in the mountains of Europe and most of the former USSR since the Second World War. One notable trend over the course of the MAB-6 programme, however, has been the increasing consideration of socio-economic-cultural systems in projects in western Europe, particularly in France and Spain. This undoubtedly reflects general trends in society and science, but would not have been possible without support from funding institutions.

Over the past decades, the societies of Europe's mountains have been increasingly influenced by external forces: tourism, agricultural subsidies and massive increases in flows of traffic and goods. The importance of these external forces has been recognized in a number of MAB-6 projects and indicates, even more than in 1973, that local peoples' needs and their ability to control their environment are crucial issues. As the Obergurgl project suggested, and the Swiss and Catalan projects have shown more rigorously, the long-term future of mountain communities requires not only ecological knowledge but also understanding of the cultural heritage and expectations of these communities. Patterns of land use are central in this and, as the Swiss MAB programme revealed, decisions regarding these patterns are driven by deep cultural, as well as economic, considerations.

Unfortunately, these socio-cultural characteristics of mountain regions are far harder to measure, and even more difficult to aggregate spatially, than the biophysical components of the landscapes that they influence. In the terminology of the Berchtesgaden project, the real driving forces of change in mountain environments are often at the 'dynamic' and 'strategic' levels; though the biophysical constraints of the 'process' level must never be ignored. It may well be that the systems modelling approach, inherited originally from the IBP, developed beyond its effective limit during the MAB-6 programme. On one hand, the Berchtesgaden models provide a conceptual means to explain the complexity of levels of driving forces that shape mountain environments. On the other hand, one may conclude that, in over-emphasizing the quantitative landscape ecology, and not the human ecology, of these environments – as evidenced particularly by the use of a GIS as a unifying framework – these models represent a retrogression, from the point attained in the Swiss programme, in the evolution of models for understanding mountain environments and their populations.

One piece of evidence for this conclusion is provided by use of the results of the German and Swiss programmes in policy formulation. The former are now being used to develop a master plan for Berchtesgaden National Park as well as guidelines for the protection and management of German biosphere reserves. In contrast, the latter have had a much wider application – not only in protected areas – providing the basis for master plans in all of the test areas, as well as policies both for agriculture and tourism at the national level and in relation to the Alpine Convention.

In conclusion, while conceptual models are necessary for research that can be effectively used for formulating policies for the long-term benefit of mountain communities, these models must consider a wide range of variables: ecological, economic, cultural, and so on. Computer technology may provide valuable assistance in storing, processing and evaluating data, and developing and illustrating possible scenarios. However, human networks are also critical; scientists must be willing to interact with, and learn from, local people and should show the relevance of their research to decision-makers at local to national levels. These conclusions are not unique to mountain environments. They are relevant to applied interdisciplinary research in general, and the next section draws on the experience of the MAB-6 programme to provide some suggestions for future research.

LESSONS FOR INTERDISCIPLINARY RESEARCH

At the 10th anniversary symposium of the MAB programme in 1981, the prime mover and co-ordinator of the Swiss MAB programme (B. Messerli, 1984: 241–2) commented: 'How are we able to think in integrated systems? How are we able ... when our educational system divides everything into disciplines? The experience of several years of Swiss work on mountains within the MAB Programme has showed that considerable efforts were necessary for the education of participating scientists. Most of us are not trained in interdisciplinary work, and it is difficult to adapt the approaches, methods and results of discipline-oriented work to the problems of an integrated and dynamic man-environment system. In too many research projects, scientists work alone, with neither the time nor energy to think of eventual research applications or the needs of local inhabitants and decision-makers. These people as well as the scientists, need new modes of thinking to help them make sound planning decisions in the face of so many considerations. The MAB Programme has a unique contribution to make in bringing about the desired changes.'

Thirteen years later, it is possible to draw many lessons from the rich experience of the diverse projects conducted within the framework of the MAB programme in the mountains of Europe and the former USSR. As noted at the beginning of the book, such information is of particular importance as an increasing number of interdisciplinary research programmes are being planned and implemented. Two groups of guidelines are presented below: for interdisciplinary research in general; and for interdisciplinary projects that are intended to be applied and effective – as the MAB-6 programme was meant to be. Many of these guidelines may appear self-evident but, as shown by the projects discussed in this book, were not obvious during either the planning or the implementation stages of MAB-6 projects.

Interdisciplinary research: general guidelines

At the beginning of any interdisciplinary research project, the participants must agree on a clear set of objectives and framework(s) for research. There are many reasons for this, apart from the obvious one of obtaining financial support. First, every scientific discipline and sub-discipline has an individual and specific language. Second, many concepts are used in contrasting ways in different disciplines and sub-disciplines. One particular example relates to scales of space and time, which vary greatly, for instance, between geomorphologists, economists and entomologists. If terminologies and frameworks – conceptual and/or data-related – that are understood by, and acceptable to, all participants are not decided at an early stage, the project is likely to result in no more than a series of studies carried out in the same region, which may be brought together in a multidisciplinary report or book but cannot be linked in a meaningful way. Third, all participants must accept their roles in the project. If some view their role primarily as providing information for others' research, with little exchange, they may have little interest in participating. Each participant should be able both to contribute and to use data and information collected by others during the project.

All of these points can be reduced to two themes: communication and direction. Clear communication between team members is essential, and may be fostered in many ways, including formal and informal meetings and presentations; sharing living-quarters; and data storage/management and simulation systems, such as GIS, with an open and understandable structure. At the same time, clear direction is required. This may come from a multidisciplinary committee, answerable to the funding agency, although the primary focus for participating scientists should be a small co-ordinating team or an individual co-ordinator. In each case, the co-ordinator(s) must clearly define each team member's responsibilities, including the completion of studies within allotted budgets and time limits. Of course, some flexibility is necessary, but only to the extent that the project's objectives are not endangered. If there is one co-ordinator, he/she must be able to concentrate on his/her very complex duties, in terms of management and understanding the component parts of the project, many of which will be from outside his/her primary discipline. Thus, co-ordinators from universities should be relieved of as many other responsibilities as possible; or a private consultant may be hired – an increasing trend.

In addition to communication and direction, two other ingredients are vital: time and money. A successful interdisciplinary project requires years to plan, implement and write up. In particular, the planning stage tends to be much longer than for other types of project, merely because of the need to develop good communication and understanding and to agree on objectives and frameworks for research. At this stage, one appropriate strategy may be to hold short but intense meetings of a few days, at which potential participants can interact and, before committing themselves to the project, determine whether they even want to be involved; a decision that may be made for them by the co-ordinator or other participants. Such brainstorming meetings, at which team members concentrate only on issues of current relevance in the project, are also invaluable during the implementation and synthesis stages.

Finally, adequate financial resources for the entire life of the project must be assured at its beginning. In the planning stage, resources for holding meetings and exploring possible approaches are necessary. In the research phase, all team members must be sure that the efforts they have contributed to developing the project will be rewarded by enough time and resources to carry out their own element of it. By the time that research concludes and synthesis begins, many team members may have started work on other projects or be 'burned out'. If full advantage is to be made of their unique experience and a useful synthesis prepared, there must be enough time and resources to bring them together and for an editor – preferably the project co-ordinator – to make a full synthesis. This happened in very few MAB-6 projects.

Effective applied interdisciplinary research

All of the guidelines presented above are relevant for applied interdisciplinary projects of the type foreseen by the MAB-6 Lillehammer group. However, as shown by the experience of the MAB-6 programme in western Europe, effective research projects require many other ingredients, which add to their complexity. Good relationships must be developed not only between the participating scientists, but also to decision-makers. As the importance of external influences on individual communities – whether in mountains or elsewhere – increases, relevant decision-makers may be not only local, but also regional, national, or even international. However, it is one thing to influence policy development; it is another to ensure that policies will be implemented and that the desired outcomes will result. Consequently, as many participating scientists as possible should also communicate and interact with local people, through: meetings and other fora for discussion; publications, using simplified terminology and graphics as appropriate; and, perhaps, the establishment of formal organizations if suitable ones do not already exist. In all cases, communication must be two-way, not merely from scientists to other people.

The intent to conduct applied research directed towards the long-term future of local communities greatly complicates the setting of objectives and research frameworks, as these need to be agreed by a far more diverse group of actors. Often, themes that are important to local people may not be clear at the beginning of the project. They may even be hidden, intentionally or unintentionally, until local people trust the project team sufficiently to voice their concerns and expectations of the project. Thus, scientific surveys of beliefs, attitudes and perceptions, and meetings should take place from an early stage. Simulations and scenarios – using computers, local experience, graphic media (including childrens' drawings) and other tools – can also be used to define the issues of greatest local concern. The challenge is then to incorporate these into the research programme and to maintain its interest for participating scientists and local people.

Finally, extra time and other resources are essential in the synthesis phase, which may take an unanticipated form because of local interest in the results of the project, and continue beyond its expected and formal end. Given adequate levels of local interest, local or regional institutions may provide necessary resources that are not available from traditional funding agencies. Optimally, people who have gained experience in the project should remain in the area after its completion; though this is often not possible for scientists who follow academic career paths.

A LEGACY FOR THE FUTURE

As Bruno Messerli anticipated in 1981, the MAB programme has made unique contributions to scientific knowledge, policies for individual mountain communities and wider mountain regions, and the lives of mountain people. New ways of thinking have evolved among many scientists and in mountain communities. Yet many challenges remain; these new ideas must continue to evolve in the highly dynamic and uncertain setting of the mountains of Europe and the former USSR.

In many of the areas where MAB-6 projects were conducted, they have an ongoing legacy in research, planning, or both. In both Davos and Grindelwald, Switzerland, the GIS developed during the MAB-6 project are still used for research, having been transferred from mainframe computers to personal computers; one of the instances of technological evolution during and since the MAB programme. In both Grindelwald and Berchtesgaden, Germany, a GIS is being used for environmental monitoring and planning. New research in Grindelwald is providing a better understanding of the environmental effects of trade-off decisions made by people who work in a mixture of economic sectors. This work builds on studies in Kenya that showed some of the shortcomings of the approaches used in the original MAB-6 research in Grindelwald. In the Catalan Pyrenees, many activities based on the findings of the MAB-6 project are taking place in the study area, often through the involvement of individuals who took part in the project. Similarly, the results of projects in the French Massif Central are being used in developing strategies for sustainable development. In all of these cases, application of the results of MAB projects has been successful because of the interactive communication built up between scientists and local people during and since the projects.

Outside the MAB-6 test areas, many of the methodologies and understanding developed during the research projects are being used in both Europe and other parts of the world. In some cases, the results of MAB projects are being used to develop policies at the national level: for instance, for guidelines for monitoring and management in biosphere reserves in Germany, or for strategic policies for agriculture and tourism in Switzerland. The GIS-based methodologies developed in Berchtesgaden are also being used in other countries, including China and Israel. The MAB-6 project in the Pays d'Enhaut, Switzerland, was particularly fruitful not only in providing an initial framework for the project in the Catalan Pyrenees, but also for applied research on public involvement in planning for mountain communities in Japan and the western United States.

The MAB-6 projects in western Europe took place in regions that, especially since the Second World War, had undergone massive economic, social and cultural transformations. The conceptual framework developed in the Swiss and German projects has already proved its worth for understanding the processes involved in these transformations in the Alps and the Spanish Catalan Pyrenees. Today, this framework, and perhaps also the experience of the project in the Spanish Catalan Pyrenees – which took place not long after the end of the Franco regime – may be of particular value for applied research in the mountains of eastern and central Europe and the former USSR, whose societies are also undergoing very rapid transformations following the demise of Communist rule.

Two areas of change of particular relevance for research within the framework of MAB – particularly using simulation approaches – may be cited. In agriculture and forestry, the reversion of property from state to private ownership, together with uncertainty in markets, means that the current situation is very fluid and long-term planning is a major challenge. In tourism, many of the traditional visitors from the region are no longer coming because they now have the money to travel to the 'west' or, conversely, do not have enough money to take holidays, which are no longer provided by employers and unions. At the same time, the number of visitors from the 'west' is increasing, with demands for facilities of different types and quality; and 'western' companies see new possibilities for investment, often welcomed by new governments hungry for hard currency.

Some of the most significant changes seem likely to take place in two of the areas where international biosphere reserves have recently been designated: on the Polish border with the Czech Republic (Giant Mountains) and with Slovakia (Tatras). Both of these regions have long histories of human use, with significant current environmental impacts from air pollution and tourism. Yet past research in these biosphere reserves, as in most others in central and eastern Europe, focused on describing their biophysical systems, with little consideration of socio-economic systems or their interactions with biophysical systems. The effective implementation of international biosphere reserves, in itself, poses great challenges. Together, the models developed in MAB-6 projects in western Europe and the concepts of biosphere reserves provide a powerful basis for defining strategies for the future of mountain regions which consider the needs of both conservation and local communities. In the current era of rapid change in the mountains of eastern and central Europe, the application of these MAB ideas may be critical. But this will require the further evolution of new ways of thinking, and enough time and resources to apply them.

BIBLIOGRAPHY

(Citations to National Committees refer to National Committees for Man and the Biosphere, unless otherwise stated)

Adamina M. (1982). *Di Bewertung der visuellen Qualität der Landscahft, dargestellt am Beispiel des Aletschtestgebiets.* Diplomarbeit, Geographisches Institut, Universität Bern.

Aegerter R. (1983) *Grindelwald: Beiträge zue Geschichte der Besiedlung und Landwirtschaft vom Mittelalter bis ins 19. Jahrhundert.* Dissertation, Geographisches Institut, Universität Bern.

Aegerter S. and P. Messerli (1981) The impact of hydroelectric power plants on a mountainous environment: a technique for assessing environmental impacts. *Fachbeiträge zur schweizerischen MAB-Information 9.* Bundesamt für Umweltschutz, Bern.

Akhaminov A.D. and P.M. Polan (eds.) (1990) *Problems of mountain industrial economy and distribution of population,* Institute of Geography, USSR Academy of Sciences, Moscow.

Alam, S.M. and A.H. Kidwai (eds.) (1987) *Regional Imperatives in Utilization and Management of Resources – India and the USSR,* Concept, Delhi.

Alcala, J. and J. Nunes (1988) Metodologia de la encuesta agraria. *Report 13, MAB 6* Alt Prinieu, La Seu D'Urgell.

Alvera, B. (1990 *Ciclos de elementos minerales en dos pinares pirenaicos.* Doctoral thesis, Universidad Complutense de Madrid.

Anglada, S. et al. (1980) La vida rural en la montaña espanola. *Monografiás del Instituto de Estudios Pirenáicos* 107.

Anon (1989) *Proceedings, Carpathian Tourist Conference.* Zeszyty Uniwersytetu Jagiellonskiego CMLVII. Krakow (mainly in Polish; some papers and summaries in English).

Apel H. (1983) Dynamische Simulation eines Bergökosystems mit einem Anhang über das Input–Output Modell (Testgebiet Grindelwald). *Schlussbericht zum schweizerischen MAB-Programm 5.* Bundesamt für Umweltschutz, Bern.

Ashdown M. and J. Schaller (1990) Geographic Information Systems and their application in MAB projects, ecosystem research and environmental monitoring. *MAB-Mitteilungen 34,* German National Committee, Bonn.

Ashdown M.L.C., J.G. Köppel and J. Schaller (eds.) (1988) Ecosystem inquiry in New Zealand and the Federal Republic of Germany. Discussion paper: Method and model testing, environmental monitoring/impact assessment and biosphere reservation. Planungsbüro Dr. Schaller, Kranzberg.

Austrian MAB programme (1983) MaB-Kartenband 'Hohe Tauern'. *Veröffentlichungen des Oesterreichischen MAB-Programmes 7.* Universitätsverlag Wagner, Innsbruck.

Avakyan G.E. (1989) *People and mountains: history, theory, practice,* Myisl, Moscow (in Russian).

Bachnuber R. et al. (1983) Landschaftsökologische Modelluntersuchung Ingolstadt (Moding)-Hauptbericht. Lehrstuhl für Landschaftsökologie, Munich-Weihenstephan Technical University.

Badenkov Y.P. (1987) Problems of the development of mountainous regions of the USSR within the framework of the project 'Mountain Ecosystems' under Unesco 'Man and the Biosphere' Programme. *In* S.M. Alam and A.H. Kidwai (eds.) *Regional Imperatives in Utilization and Management of Resources – India and the USSR*. Concept, Delhi: 419–26.

Badenkov Y.P. (1990) Sustainable development of the mountain regions of the USSR. The realities, the role of science, and research orientations. *Mountain Research and Development* 10: 129–39.

Badenkov Y.P. (1992) Mountains of the former Soviet Union: value, diversity, uncertainty *In* P.B. Stone (ed.) *The State of the World's Mountains*, Zed Books, London: 257–97.

Badenkov Y.P. et al. (1992) Caucasia *In* Turner B.L. et al. (eds.) *The Earth as Transformed by Human Action*, Cambridge University Press, Cambridge: 513–31.

Balcells E. (1976a) Observaciones sobre el ciclo biológico de anfibios de alt montaña y su interés en la detección del inicio de la estimación vegetativa. *Publicaciones del Centre Pirenaico de Biologia Experimental* 7: 55–154.

Balcells E. (1976b) *El Pirineo: contraste de paisajes: enlace de pueblos*. Comisión Internacional de los Pirineos, Ministerios de Asuntos Exteriores, Madrid.

Balcells E. (1983) Estado actual del Proyecto M.A.B. num. 510: 'Estudio multidisciplinar e integrado de pinares pirénaicos.' *Pirineos* 120: 65–87.

Balcells E. (1984) Estudio comparado de las cuencas altas del Subordán y del Veral y de las unidades étnicas que utilizan sus recursos. *Pirineos* 123: 5–152.

Bätzing W. (1988) Ökologische Labilität und Stabilität der alpinen Kulturlandschaft. *Fachbeiträge zur schweizerischen MAB-Information 27*. Bundesamt für Umweltschutz, Bern.

Batisse M. (1986) Developing and focusing the biosphere reserve concept. *Nature and Resources* 22 (3): 2–11.

Beguin C. and J.-P. Theurillat (1982) Analyse de la végétation et du paysage de la région d'Aletsch. *Fachbeiträge zu schweizerischen MAB-Information 11*. Bundesamt für Umweltschutz, Bern.

Beguin C. and J.-P. Theurillat (1987) Analyse de la végégatation et du paysage d'Aletsch MAB-6: Valeurs de protection et charges écologiques. *Rapport final FNS 4.331*, Bundesamt für Umweltschutz, Bern.

Beissmann H. (1989) Plausibilitätsanalysen mit Hilfe eines EDV-gestützten themakartographischen Informationssystems. Berichte und Informationen 14. Österreichische Akademie der Wissenschaften, Institut für Kartographie, Vienna.

Bellwald S. and H. Graf (1985) Der Walt im Aletschgebiet – Zustand und Entwicklungstendenzen. *Schlussbericht zum schweizerischen MAB-Programme 17*. Bundesamt für Umweltschutz, Bern.

Belochova I. (1982) *Vyberova bibliografie Krkonoš 1963–1978*. Sprava Krkonošskeho narodniho parku, Vrchlabi.

Beniston M. and M.F. Price (1991) Climate scenarios for Alpine regions: a collaborative effort between ICALPE and ProClim. *Environmental Conservation* 18(4): 360–3.

Binz H.R. and O. Wildi (1988) Das Simulationsmodell MAB-Davos. *Schlussbericht zum schweizerischen MAB-Programm 33*. Bundesamt für Umweltschutz, Bern.

Breymeyer A. et al. (1994) *Biosphere reserves in Poland*. Polish National MAB Committee, Warsaw.

Briand F. et al. (1989) *The Alps*. ICALPE, Le Bourget-du-Lac, (also published in French and Italian in 1992).

Brugger E.A. et al. (eds.) (1984a) *Umbruch im Berggebiet*, Haupt, Bern.

Brugger E.A. et al. (eds.) (1984b) *The Transformation of Swiss Mountain Regions*, Haupt, Bern.

Bulgarian National Committee (1983) *Proceedings of the conference on high-mountain ecosystems*, Vratsa, Bulgaria, 24–29 October 1983, 3 volumes, Sofia (in Bulgarian, English summaries).

Campillo X. and X. Sanclimens (1986) Viatges i visions dels temps antics. *Report 3, MAB 6* Alt Pirineu, La Seu d'Urgell.

Campillo X. et al. (1985) Programa MAB 6 Alt Pirineu: Introducción. *Report 1, MAB 6* Alt Pirineu, La Seu d'Urgell.

Campillo X. et al. (1987) La población: evolución reciente, situación actual y prospectiva para el año 2000. *Report 8, MAB 6*, Alt Pirineu, La Seu d'Urgell.

Campillo X. et al. (eds.) (1993) *MAB 6 Alt Pirineu: Sintesis del programa*. Ministerio de Obras Públicas y Transportes, Madrid.
Cernusca A. (ed.) (1977) Alpine Grasheide Hohe Tauern. *Veröffentlichungen des Oesterreichischen MaB-Programms 1*. Universitäts Wagner, Innsbruck.
Cernusca A. (ed.) (1978) Oekologische Analysen von Almflächen im Gasteiner Tal. *Veröffentlichungen des Osterreichischen MaB-Programms 2*. Universitätsverlag Wagner, Innsbruck.
Cernusca A. (ed.) (1989a) Struktur und Funktion von Graslandökosystemen im Nationalpark Hohe Tauern. *Veröfftenlichungen des Oesterreichischen MaB-Programmes 13*. Universitätsverlag Wagner, Innsbruck.
Cernusca A. (1989b) Ecosystem research in Austrian Central Alps (Hohe Tauern) *In* A. Cernusca (ed.) *Struktur und Funktion von Graslandökosystemen im Nationalpark Hohe Tauern. Veröffentlichungen des Oesterreichischen MaB-Programmes 13*. Universitätsverlag Wagner, Innsbruck: 587–625.
Cernusca A. (1991) Ecosystem research on grassland in the Austrian Alps and in the Central Caucasus. *In* Esser G. and D. Overdieck (eds.) *Modern Ecology: Basic and Applied Aspects* Elsevier, Amsterdam: 233–71.
Chapalay A. (1984) The viewpoint of the local population of the Pays d'Enhaut region *In* F. di Castri, F.W.G. Baker and M. Hadley (eds.) *Ecology in Practice*, vol. 2. Tycooly, Dublin: 295–7
Chocarro C. (1990) *Estudios ecológicos sobre los prados de siega del Pirineo central español: composición floristica, producción y calidad*. Doctoral thesis, Universidad de Navarra, Pamplona.
Chocarro C. et al. (1987) Meadows of Central Pyrenees: Floristical composition and quality. *Pirineos* 129: 7–33.
Chocarro C. et al. (1990) Historical evolution of natural resource use in the central Pyrenees of Spain. *Mountain Research and Development* 10: 257–65.
Chupakhin V.M. (1987) *Vertical zonation of geosystems of Central Asia and Kazakhstan*. Nauka, Alma-Ata (in Russian).
Chupakhin V.M. and B.I. Kochurov (eds.) (1982) *Mountain geosystems*. Soviet Committee for MAB and USSR Geographical Society, Moscow and Alma-Ata (in Russian).
Coendet M. (1979) *Thermische Gliederung und lufthygienische Probleme im Talkessel von Grindelwald*. Diplomarbeit, Geographisches Institut, Universität Bern.
Combe J. and C. Frei (1986) Die Bewirtschaftung des Bergwaldes. *Schlussbericht zum schweizerischen MAB-Programm 22*. Bundesamt für Umweltschutz, Bern.
Corominas J. and J. Puig (1988) Recursos y necesidades energeticas. *Report 17, MAB 6* Alt Pirineu, La Seu d'Urgell.
Council of Europe (1978) *Report of the seminar on pressures and regional planning problems in mountain regions*. Delegate for Regional Planning, Bern.
Creus J. (1983) El clima del Alto Aragón occidental. *Monografias del Instituto de Estudios Pirenaicos* 109.
Czechoslovak National Committee (1990) *Report* (in Czech). Prague.
Darbeliay C. and E. Stucki (1984) Glimpses from the MAB-Pays d'Enhaut project. *In* F. di Castri, F.W.G. Baker and M. Hadley (eds.) *Ecology in Practice*, vol. 2. Tycooly, Dublin: 283–94.
di Castri F., F.W.G. Baker and M. Hadley (eds.) (1984) *Ecology in Practice*, 2 volumes, Tycooly, Dublin.
Direction du Project MAB-Pays d'Enhaut (DPMPE) (1985a) Agriculture Pays d'Enhaut. *Schlussbericht zum schweizerischen MAB-Programm 9*. Bundesamt für Umweltschutz, Bern.
Direction du Project MAB-Pays d'Enhaut (DPMPE) (1985b) Population et économie régionale. *Schlussbericht zum schweizerischen MAB-Programm 14*. Bundesamt für Umweltschutz, Bern.
Direction du Projet MAB-Pays d'Enhaut (DPMPE) (1985c) Tourisme Pays d'Enhaut. *Schlussbericht zum schweizerischen MAB-Programme 15*. Bundesamt für Umweltschutz, Bern.
Direction du Projet MAB-Pays d'Enhaut (DPMPE) (1986) Aménagement du territorie et protection de la nature au Pays d'Enhaut. *Schlussbericht zum schweizerischen MAB-Programm 26*. Bundesamt für Umweltschutz, Bern.
Direction du Projet MAB-Pays d'Enhaut (DPMPE) (1988a) Foresterie et économie du bois

au Pays d'Enhaut. *Schlussbericht zum schweizerischen MAB-Programm 32*. Bundesamt für Umweltschutz, Bern.

Direction du Projet MAB-Pays d'Enhaut (DPMPE) (1988b) L'homme et la nature au Pays d'Enhaut (2 vols.). *Schlussbericht zum schweizerischen MAB-Programm 36*. Bundesamt für Umweltschutz, Bern.

Dobesch H. (ed.) (1983) Klimatologische Untersuchungen in den Hohen Tauern 1974–1980. *Veröffentlichungen des Oesterreichischen MaB-Programms 6*. Universitätsverlag Wagner, Innsbruck.

D'Oleire-Oltmanns W. (1985) Das Programm MAB 6 in der Bundesrepublik: geplante und durchgefürte Untersuchungen im Nationalpark Berchtesgaden *In Ministère de l'Environnement, Bilans et perspectives de recherches scientifiques pour une meilleure gestion des milieux montagnards*. Collection Recherche Environnement 25, SRETIE, Ministère de l'Environnement, Paris: 33–43.

Dolishniy M.I. (ed.) (1988) *Ukrainian Carpathians: Economy*, Naukova Dumka, Kiev (in Russian).

Dollfus O. (ed.) (1981) *Recherches en Briançonnais*. CEMAGREF, Grenoble.

Dvořak J. and V. Pilous (1988) General index of the yearbook Opera Concortica for volume 1–25 (1964–1988). *Opera Concortica* 25: 143–61.

Dzhaoshvili V.S. et al. (1988) Mountain areas: rational land use and the distribution of population and land use, *Geography of the USSR* Vol. 18, All-Union Institute for Scientific and Technical Information, Moscow (in Russian).

Farrero A. and M.A. Villamuera (1987) Los planes de acción contra los incendios forestales. *Report 7, MAB 6*, Alt Pirineu, La Seu d'Urgell.

Fillat F. (1980) *De la trashumancia a las nuevas formas de ganadera extensiva: Estudio de los valles de Ansó, Hecho y Benasque*. Doctoral thesis, Universidad Politécnica de Madrid.

Fischesser B. and M. Dubost (eds.) (1990) *Les formations paturées d'altitudes*. CEMAGREF, Grenoble.

Fracheboud J.-F. (1984) Modèle de simulation d'un écosystème de montagne, aire-test Pays d'Enhaut. *Schlussbericht zum schweizerischen MAB-Programm 6*. Bundesamt für Umweltschutz, Bern.

Franch J. (1985) *Estudio del proceso de descomposición de tocones de* Pinus silvestris *en el bosque de San Juan de la Peña (Jaca)*. Doctoral thesis, Universidad de Barcelona.

Frank F. (1991) *Davos-MaB-Magazin*. Gemeinde Davos.

Franz H. (ed.) (1980) Untersuchungen an alpinen Boden in den Hohen Tauern 1974–1978. Stoffdynamik und Wasserhaushalt. *Veröffentlichungen des Oesterreichischen MaB-Programms 3*. Universitätsverlag Wagner, Innsbruck.

Franz H. (ed.) (1981) Bodenbiologische Untersuchungen in den Hohen Tauern 1974–1978. *Veröffentlichungen des Oesterreichischen MaB-Programms 4*. Universitätsverlag Wagner, Innsbruck.

Franz H. (ed.) (1985) Beiträge zu den Wechselbeziehungen zwischen den Hochgebirgsökosystemen und dem Menschen. *Veröffentlichungen des Oesterreischischen MaB-Programms 9*. Universitätsverlag Wagner, Innsbruck.

Freiburghaus D. and W. Zimmermann (1984) Fallstudie 'MAB (Man and Biosphere),' *Informationsbulletin der Programmleitung NFP 6, No. 11*. Forschungsstelle Wissenschaft und Politik, Bern.

French National Committee (1984) La contribution française au programme international 'L'Homme et la Biosphère.'*Collection Internationale 1, Mission Environnement-Développement*, Ministère de l'Environnement, Paris.

French National Committee (1986) La contribution française au programme international 'L'Homme et la Biosphère.' *Collection International 2, Mission Environnement-Développement*, Ministère de l'Environnement, Paris.

French National Committee (1988) Rapport sur les activités de Comité Français du MAB. Période 1986–1988. *Collection Recherches Internationales 33, Mission Environnement-Développement*, Secrétariat d'Etat auprès du Premier Ministre, chargé de l'Environnement, Paris.

French National Committee (1990) Rapport sur les activités de Comité Français du MAB, Période: 1988–1990. *Collection Recherches Internationales 35, Mission Environnement-*

Développement, Secrétariat d'Etat auprès du Premier Ministre, chargé de l'Environnement et de la Prévention des Risques Technologiques et Naturels Majeurs, Paris.

French National Committee (1992) *Rapport sur les activités de Comité Français du MAB, Période: 1990–1992*. Direction de la Nature et des Paysages, Ministère de l'Environnement, Paris.

Ganyet J. and A. Tulla (1988) The high Pyrenees MAB-6 program: results and conclusions. *Documents d'Anàlisi Geogràfica* 12: 179–86 (in Catalan).

García-González R. (1987) Estudio del crecimiento postnatal en corderos de raza Rasa Aragonesa ecotipo Ansotano. *Colleción de Estudios Altoaragoneses 13*, Instituto de Estudios Altoaragoneses, Huesca.

García-González R. et al. (1990) Patterns of time and space use by livestock in the Pyrenean summer ranges: a case study in the Aragon valley. *Mountain Research and Development* 10: 241–55.

García-Ruiz J.M. (1976) Modos de vida y niveles de renta en el Prepirineo del Alto Aragón occidental. *Monografias del Instituto de Estudios Pirenaicos* 106.

García-Ruiz J.M. and T. Lasanta-Martinez (1990) Land-use changes in the Spanish Pyrenees. *Mountain Research and Development* 10: 267-79.

García-Ruiz J.M. et al. (1986) La acumulacíon de la nieve en el Pirineo Central y su influencia hydráulica. *Pirineos* 127: 27–72.

Gensac P. (ed.) (1985) *L'amenagement de la haute montagne et ses consequences sur l'environnement: le canton d'Aime (Savoie). Rapport de première phase*. Université de Savoie, Chambéry.

Gerasimov I.P., I. Bravar and J. Dresh (eds.) (1980) *Alpes-Caucase: Problemes actuels de la Géographie constructive des pays de montagne*, Naouka, Moscow (in Russian).

Gerasimov I.P. and I. Galabov (eds.) (1984) *The Greater Caucasus – The Stara Planina (the Balkans)*, Naouka, Moscow (in Russian).

German National Committee (1979) Internationales Seminar 'Schutz und Erforschung alpiner Oekosysteme.' *MAB-Mitteilungen 4*. German National Committee, Bonn.

German National Committee (1982) Bericht über das internationale MAB-6 Seminar 'Der Einfluss des Menschen auf Hochgebirgsökosysteme im Alpen- und Nationalpark Berchtesgaden.' *MAB-Mitteilungen 9*, German National Committee, Bonn.

German National Committee (1983a) Ziele, Fragestellungen und Methoden: Oekosystemforschung Berchtesgaden. *MAB-Mitteilungen 16*, German National Committee, Bonn.

German National Committee (1985a) Proeedings of the III. International MAB-6 Symposium 'The impact of human activities on high mountain ecosystems of the alpine and national park Berchtesgaden.' *MAB-Mitteilungen 19*, German National Committee, Bonn.

German National Committee (1985b) Bericht über das IV. international Seminar 'Der Einfluss des Menschen auf Hochgebirgsökosysteme im Alpen- und Nationalpark Berchtesgaden.' *MAB-Mitteilungen 21*, German National Committee, Bonn.

Gigon A. (1981) Ökologische Stabilität: Typologie und Realisierung. *Fachbeiträge zur schweizerischen MAB-Information 7*. Bundesamt für Umweltschutz, Bern.

Glaser G. (1978) The present status and future strategies of Unesco's Man and the Biosphere (MAB) programme. *GeoJournal* 2: 488–91.

Golubets M.A. et al. (eds.) (1988) *Ukrainian Carpathians: Nature*. Naukova Dumka, Kiev (in Russian).

Gómez D. (1989) *Flora y vegetación de Peña Monteñesa-Sierra Ferrera y Valle de la Fueva (Alto Sobrarbe, Huesca)*, Universidad de Barcelona.

Goshko Y.G. et al. (1989) (eds.) *Ukrainian Carpathians: Culture*. Naukova Dumka, Kiev (in Russian).

Grabherr G. et al. (1992) Hemerobie Österreicheischer Waldökosysteme-Zwischenbericht. Institut fur Planzenphysiologie, University of Vienna.

Grodzińska K. and R. Olazcek (eds.) (1985) *National Parks in danger*. Polish Scientific Publishers, Warsaw (in Polish, some English summaries).

Grosjean G. et al. (1986) Ästhetische Bewertung ländlicher Räume. *Schlussbericht zum schweizerischen MAB-Programm 20*. Bundesamt für Umweltschutz, Bern.

Guet J. (1974) Le Briançonnais *In Propositions de la délégation française à la troisième réunion de Comité Internationale de Coordination du Programme MAB, Paris*: 88–92.

Günter T.F. (1985) Landnutzungsänderungen in einem alpinen Tourismusort. *Schlussbericht zum schweizerischen MAB-Programm 13*. Bundesamt für Umweltschutz, Bern.

Günter T.F. (1987) Natur- und Landschaftsschutz als Element der qualitativen Fremdenverkehrsentwicklung. *Fachbeiträge zur schweizerischen MAB-Information 26*. Bundesamt für Umweltschutz, Bern.

Gurrera J. (1988) El transporte publico de viajeros. *Report 14, MAB 6* Alt Pirineu, La Seu d'Urgell.

Gurrera J. et al. (1987) La red viaria. *Report 6, MAB 6* Alt Pirineu, La Seu d'Urgell.

Gvishiani D.M. and V.E. Sokolov (eds.) (1979) *Programme on Man and the Biosphere in Socialist Countries*. USSR Committee for Man and the Biosphere, Moscow.

Gvishiani D.M. and V.E. Sokolov (eds.) (1981) *Integration of sciences in the programme on Man and the Biosphere*. USSR Committee for Man and the Biosphere, Moscow.

Haber W. (ed.) (1986) Mögliche Auswirkungen der geplanten Olympischen Winterspiele 1992 auf das regionale System Berchtesgaden. *MAB-Mitteilungen 22*, German National Committee, Bonn.

Haber W. (1990) Using landscape ecology in planning and management. *In* I.S. Zonneveld and R.T.T. Forman (eds.) *Changing Landscape: an Ecological Perspective*. Springer, New York: 217–23.

Haber W., L. Spandau and K. Tobias (eds.) (1990) Ökosystemforschung Berchtesgaden, Schlussbericht über die Arbeiten der Fachdisziplinen. *Texte 15/90*, Umweltbundesamt, Berlin.

Hadac E. and J. Terray (eds.) (1991) *Flora of the Bukovske Vrchy* (in Slovak). Priroda, Bratislava.

Hefti R. and U. Bühler (1986) Zustand und Gefährdung der Davoser Waldungen. *Schlussbericht zum schweizerischen MAB-Programme 23*. Bundesamt für Umweltschutz, Bern.

Himamowa B. (1974) *The Obergurgl model: a microcosm of economic growth in relation to limited ecological resources*. International Institute for Applied Systems Analysis, Laxenburg.

Hoppler F. et al. (1985) Koordination touristischer und verkehrlicher Erschliessung (Testgebiet Grindelwald), *Schlussbericht, MAB-Projekt 4.188*. Institut für Verkehrsplanung und Transporttechnik, ETH-Zürich.

ICALPE (1992) *Montagnes et forêts méditerranéennes*. ICALP, Le Bourget-du-Lac.

Ives J.D. and B. Messerli (1990) Progress in theoretical and applied mountain research, 1973–1989, and major future needs. *Mountain Research and Development* 10: 101–27.

Jahn A. (1985) *Karkonosze Polskie*. Ossolineum, Wroclaw.

Jenik J. (1983) Succession on the Polonina balds in the western Bieszczady, the eastern Carpathians. *Tuexemia* 3: 207–16.

Jenik J. (1987) Selection and role of biosphere reserves in Czechoslovakia *In Czechoslovak National MAB Committee, Proceedings: European MAB Conference on Biosphere Reserves and Ecological Monitoring*. Ceske Budejovice: 23–32.

Jenik J. and M.F. Price (eds.) (1994) *Biosphere reserves on the Crossroads of Central Europe*. Empora (Prague), Prague.

Jumayev T. (1989) *Mountains of Uzbekistan: Nature, economy, and recreation*. Mechnat, Tashkent (in Russian).

Kerner H.F. and L. Spandau (eds.) (1990) Sozioökonomische Auswirkungen und Szenarien zur Berglandwirtschaft im Alpenpark Berchtesgaden. *Lehrstuhl für Landschaftsökologie der Technischen Universität München-Weihenstephan.*

Kerner H.F., L. Spandau and J.G. Köppel (eds.) (1991) Methoden zur angewandten Oeosystemforschung entwickelt im MAB-Projekt 6 'Oekosystemforschung Berchtesgaden.' *MAB-Mitteilungen 35.1 and 35.2* (2 vols.). German National Committee, Freising-Weihenstephan.

Klimek K. (ed.) (1987) *Tsarichina: Biosphere reserve in Stara Planina*, Nature and Natural Resources Protection Research Centre, Polish Academy of Sciences, Krakow (in Russian, English summaries).

Klimek K. (ed.) (1989) Ochrona Przyrody i krajobrazu Karpat Polskich. *Studia Naturae* B 33.

Klimo E. (1981) Ecological consequences of the intensive mechanization and clear cutting in pure spruce stands on sites of the middle altitude. *MAB Newsletter for Project 2* No. 4: 16–18.

Klimo E. (1990) Changes of the forest site and soil processes in conditions of anthropogenic load. *MAB Newsletter for Project 2* No. 8: 17–20.

Kneubühl U. (1987) Die Entwicklungssteuerung in einem Tourismusort, untersucht am Beispiel Davos. *Geographica Bernensia P12*, Geographical Institute, University of Bern.

Kobakhidze A.D. (1985) *Social-ecological and economic problems of mountain regions of the Caucasus*, Metsniereba, Tbilisi (in Russian).

Konca B. and T. Bugaj (eds.) (1991) Badania naukowe w Karkonoskim Parku Narodowym. *Prace Karkonoskiego Towarzystwa Naukowego* 53, Jelenia Gora.

Koncek M. (ed.) (1974) *The climate of the Tatras*, Veda, Bratislava (in Slovak, English summaries).

Körner C., G. Wieser and A. Cernusca (1989) Der Wasserhaushault waldfreier Gebiete in den Oesterreichischer Alpen zwischen 600 und 2,600 m Höhe *In* A. Cernusca (ed.) Struktur und Funktion von Graslandökosystemen im National Park Hohe Tauern, *Veröffentlichungen des Österreichischen MaB Programmes Band 13*, pp. 119–53.

Koshkariov A. et al. (1994) Towards resolving the problems of regional development in the mountains of the Commonwealth of Independent States using geographic information systems *In* M.F. Price and D.I. Heywood (eds.) *Mountain Environments and Geographic Information Systems*, Taylor and Francis, London, pp. 77–97.

Kotarba A. (ed.) (1992) Special issue on Environmental transformation and human impact in the Polish Tatra. *Mountain Research and Development* 12(2).

Kotlyakov V.M. and A.V. Yashina (eds.) (1987) *Transformation of mountain ecosystems of the Greater Caucasus*, Institute of Geography, USSR Academy of Sciences, Moscow (in Russian).

Krause M. (1982) *Vergleich von Böden verschiedener Vegetations- und Nutzungsarten von der montanen bis zur oberen subalpinen Stufe.* Diplomarbeit, Geographisches Institut, Universität Bern.

Krause M. (1986) Die Böden von Davos: Ertragspotential, Belastbarkeit und Gefährdung durch Nutzungsänderungen. *Schlussbericht zum schweizerischen MAB-Programm 18.* Bundesamt für Umweltschutz, Bern.

Krippendorf J. (1986) *Alpsegen, Alptraum. Für eine Tourismus-Entwicklung im Einklang mit Mensch und Natur.* Kümmerly & Frey, Bern.

Küchli C., R. Guggenbühl and P. Messerli (1983) Öffentlichkeitsarbeit im MAB-Programm. Fachbeiträge zur schweizerischen MAB-Information 16. Bundesamt für Umweltschutz, Bern.

Kunz S. (1983) Anwendungsorientierte Kartierung der Besonnung im regionalen Masstab (Beispiel Grindelwald). *Geographica Bernensia G19*, Geographisches Institut, Universität Bern.

Kupfer F. and H. Langenegger (1985) Der Wald in Grindelwald: Zustand, Produktion und Nutzungsmöglichkeiten. *Schlussbericht zum schweizerischen MAB-Programm 11.* Bundesamt für Umweltschutz, Bern.

Kuzma J. (1989) Bibliografia Karkonoszy Polskich. *Prace Karkonoskiego Towarzystwa Naukowego 42*, Jelenia Gora.

Kyburz R. and B. Schmid (1983) Ein Gesamtmodell für die Region Pays d'Enhaut. *Schlussbericht zum schweizerischen MAB-Programm 4.* Bundesamt für Umweltschutz, Bern.

Langenegger H. and C. von Grünigen (1983) Tiefschneeskifahren und Waldverjüngung im Bereich der oberen Waldgrenze. *Fachbeiträge zur schweizerischen MAB-Information 15*, Bundesamt für Umweltschutz, Bern.

Leibundgut C. (1987) Hydroökologische Untersuchungen in einem alpinen Einzugsgebiet (Testgebiet Grindelwald). *Schlussbericht zum schweizerischen MAB-Programm 30.* Bundesamt für Umweltschutz, Bern.

Lieberherr-Gardiol F. (1984) Des hommes et des natures. Le Pays d'Enhaut en 1984 (2 vols). *Schlussbericht zum schweizerischen MAB-Programm 8.* Bundesamt für Umweltschutz, Bern.

Lieberherr-Gardiol F. and E. Stucki (1987) *Sur nos monts quand la nature* ... CERME, Château d'Oex.

Liniger H.-P. (1983) *Veränderung des Bodens im Aletschgebiet (VS) durch die traditionelle Wiesenbewässerung.* Diplomarbeit, Geographisches Institut, Universität Bern.

Malysheva L. and V. Potapenko (1993) The substantiation of the monitoring structure in the Ukraine's biosphere reserve. Paper presented at *International Conference on Mountain*

National Parks and Biosphere Reserves: Monitoring and Management, Krkonoše National Park, Czech Republic, September 1993.

Mamytov A.M. (1987). *Soils of the mountains of Middle Asia and South Kazahstan*. Ilim, Frunze (in Russian).

Mamytov A.M. et al. (eds.) (1982) *Problems of mountain development*. Ilim, Frunze (in Russian).

Marcek A. (1984) Selective bibliography on the Tatra National Park on the 35th anniversary of its legalization (period 1962–1983). *Zbornik Prac o Tatranskom Narodnom Parku* 25: 163–324.

Martinez S. (1988) Evolución de los usos del suelo 1957–1987: inventario de resultados y análisis. *Report 18, MAB 6* Alt Pirineu, La Seu d'Urgell.

Martínez-Rica J.P. (1979a) Los anfibios del Alto Aragón: un essayo de corologia. *Publicaciones del Centre Pirenaico de Biologia Experimental* 10: 7–47.

Martínez-Rica J.P. (1979b) Los reptiles del Alto Aragón. *Publicaciones del Centre Pirenaico de Biologia Experimental* 10: 49–101.

Mateu X. (1988) La administración local y los servicios municipales. *Report 16, MAB 6* Alt Pirineu, La Seu d'Urgell.

Mattig F. and H.-P. Zeiter (1984) Der touristische Wachstumsprozess im MAB-Testgebiet Aletsch, seine räumliche Ausprägung und seine Auswirkungen aud Bevölkerung, *Arbeitsmarkt und Gemeindefinanzen*. Druck AG, Fiesch.

Mayer H. (1990) Skipistenökologische Umweltverträglichkeitsprüfung der Wald-Abfahrten in Gasteiner Ski-Zirkus. *Veröffentlichungen des Osterreichischen MaB-Programms 16*. Universitätsverlag Wagner, Innsbruck.

Maza A. and R. Aspinach (1987) Fomento del sector forestal. *Report 5, MAB 6* Alt Pirineu, La Seu d'Urgell.

Messen H. and P. Luder (1987) Landschaftskonzept Aletsch. *Fachbeiträge zur schweizerischen MAB-Information 25*. Bundesamt für Umweltschutz, Bern.

Mercier A. et al. (1987) Quelques aspects de la production fourragère dans le Pays d'Enhaut. *Schlussbericht zum schweizerischen MAB-Programm 29*. Bundesamt für Umweltschutz, Bern.

Messerli B. (1984) Critical research problems in mountain ecosystems (Africa, Alps, Himalayas) *In* F. di Castri, F.W.G. Baker and M. Hadley (eds.) *Ecology in Practice*, vol. 1. Tycooly, Dublin: 234–42.

Messerli B. and J.D. Ives (eds.) (1984) *Mountain ecosystems: stability and instability*. International Mountain Society, Boulder (originally published as *Mountain Research and Development* Vol. 3(2) and 3(3), 1983, and Vol. 4(1), 1984).

Messerli B. and P. Messerli (1978) Wirtschaftliche Entwicklung und ökologische Belastbarkeit im Berggebiet (MAB Schweiz). *Geographica Helvetica* 4: 203–10.

Messerli P. (1984) Design and implementation of Swiss research on integrated mountain management *In* F. di Castri, F.W.G. Baker and M. Hadley (eds.) *Ecology in Practice*, vol. 2. Tycooly, Dublin: 277–83.

Messerli P. (1986) Modelle und Methoden zur Analyse der Mensch-Umweit-Beziehungen im alpinen Lebens- und Erholungsraum. *Schlussbericht zum schweitzerischen MAB-Programm 25*. Bundesamt für Umweltschutz, Bern.

Messerli P. (1987) The development of tourism in the Swiss Alps: economic, societal and environmental effects. *Mountain Research and Development* 7(1): pp. 13–23.

Messerli P. (1989) *Mensch und Natur im alpinen Lebensraum: Risiken, Chance, Perspektiven*. Haupt, Bern.

Messerli P. and E. Stucki (eds.) (1983) Colloque international MAB-6 les Alpes: modele et synthese. *Fachbeiträge zur schweizerischen MAB-Information 19*. Bundesamt für Umweltschutz, Bern.

Messerli P. et al. (1980) Socio-economic development and ecological capacity in a mountainous region: a study of the Aletsch-region (Switzerland). *Geographica Helvetica* 35(5): 153–60.

Messerli P. et al. (1986) Modellstudie Grindelwald zur Umweltverträglichkeit Olympischer Winterspiele im Berner Oberland 1996/2000. *Schlussbericht zum schweizerischen MAB-Programm 28*. Bundesamt für Umweltschutz, Bern.

Meyrat-Schlee E. (1983) Werte und Verhalten. Bedeutung und Wirkungsweise von Wertsystemen im Entwicklungsprozess einer Berggemeide, aufgezeigt am Besipiel von

Grindelwald. *Schlussbericht zum schweizerischen MAB-Programm 2.* Bundesamt für Umweltschutz, Bern.

Mill P. and N. Valls (1988a) La agricultura: rentas agrarias. *Report 11, MAB 6* Alt Pirineu, La Seu d'Urgell.

Mill P. and N. Valls (1988b) La ganadería. *Report 12, MAB 6* Alt Pirineu, La Seu d'Urgell.

Ministère de l'Environnement (1985) Bilans et perspectives de recherches scientifiques pour une meilleure gestion des milieux montagnards. *Collection Recherche Environnement 25*, SRETIE, Ministère de l'Environnement, Paris.

Mirek Z. (1992) Threats to the natural environment of the Tatra National Park, *Mountain Research and Development* 12 (2), pp. 193–203.

Mishev K. et al. (eds.) (1989) Natural and economic potential of the mountains in Bulgaria, Volume I, *Nature and resources*, Jusautor, Sofia (in Bulgarian).

Montserrat P. and F. Fillat (1990) The systems of grassland management in Spain *In* A. Breymeyer (ed.) *Managed Grasslands*, Elsevier, Amsterdam: 37–70.

Moser H. (1985) Die lufthygienischen Verhältnisse im Hochtal von Davos. *Schlussbericht zum schweizerischen MAB-Programm 16.* Bundesamt für Umweltschutz, Bern.

Moser P. and Moser W. (1986) Reflections on the MAB-6 Obergurgl project and tourism in an Alpine environment. *Mountain Research and Development* 6: 101–18.

Moser W. (1987) Chronik von MaB-6 Obergurgl *In* G. Patzelt (ed.) *MaB-Projekt Obergurgl, Veröffentlichungen des Oesterreichischen MaB-Programms 10.* Universitätsverlag Wagner, Innsbruck: 7–24.

Moser W. and J. Peterson (1981) Limits to Obergurgl's growth. *Ambio* 10: 68–72.

Msika B. and B. Hubert (1989) *Activités agricoles et systeme agraire dans la Vallée des Duyes (Alpes de Haute Provence): Bilan d'étape (1984–88).* INRA, Avignon.

Müller J.P. et al. (1988) Die Lebensbedingungen ausgewälter Wildterarten im MAB-Testgebiet Davos. *Schlussbericht zum schweizerischen MAB-Programm 34.* Bundesamt für Umweltschutz, Bern.

Nachbur B. (1983) *Flächennutzung und Entwicklungstendenzen der Landwirtschaftsbetriebe in ausgewählten Gemeinden des Aletschtestgebietes.*Diplomarbeit, Geographisches Institut, Universität Bern.

Naegeli-Oertle R. (1986) Die Berglandwirtschaft und Alpwirtschaft in Grindelwald. *Schlussbericht zum schweizerischen MAB-Programm 21.* Bundesamt für Umweltschutz, Bern.

Nedialkov S. (1986) *Biosphere reserves in Bulgaria*, Jusautor, Sofia (in Bulgarian).

Nedialkov S. et al. (eds.) (1985) *Proceedings, International symposium on conservation of natural areas and of the genetic materials they contain*, 3 volumes, Jusautor, Sofia (in Bulgarian, English summaries).

Nohl W. and K.D. Neumann (1986) Landschaftsbildbewertung im Alpenpark Berchtesgaden. *MAB-Mitteilungen 23*, German National Committee, Bonn.

Nosek J. (ed.) (1986) *The soil fauna of the Little Carpathians.* Slovak Academy of Sciences, Bratislava.

Observatoire Causses-Cévennes (1988) *Etre eleveur sur un causse: Le Méjan.* Parc National des Cévennes, Florac.

Observatoire Causses-Cévennes (1989) Etre eleveur sur un causse: Le Méjan. *Annales du Parc National des Cévennes 4.*

Observatoire Causses-Cévennes (1992) A la recherche de l'Aigoual perdu. *Annales du Parc National des Cévennes 5.*

Osmonov A.O. et al. (eds.) (1991) *Rational land use in mountain areas.* State University of Kirghizia, Bishkek (in Russian).

Ozenda P. and J.-L. Borel (1991) *Les conséquences écologiques possibles des changements climatiques dans l'Arc alpin.* ICALPE, Le Bourget-du-Lac.

Palanca A. (1987) Aspectos faunísticos y ecologicos de lepidópteros altoaragoneses. *Monografias del Instituto Pirenaico de Ecología 1.*

Partyka J. (1977) *Bibliografia Bieszczadska za lata 1800–1975.* Bieszczady Park Narodowy, Krakow.

Patzelt G. (ed.) (1987) MaB-Projekt Obergurgl. *Veröffentlichungen des Oesterrechischen MaB-Programms 10.* Universitätsverlag Wagner, Innsbruck.

Pawlowski B. (1956) *Flora Tatrorum.* Polish Scientific Publishers, Warsaw (in Polish).

Pedrocchi C. (1987) Fauna ornitica del Alto Aragón occidental. *Monografias del Instituto Pirenaico de Ecología 2.*

Pedrocchi C. et al. (1985) les artrópodos epigeos del Macizo de San Juan de la Peña. *Pirineos* 124 and 126.

Pfister H. (1984) Grünlandgesellschaften, Pflanzenstandort und futterbauliche Nutzungsvarianten im montan-subalpinen Bereich (Testgebiet Grindelwald). *Schlussbericht zum schweizerischen MAB-Programm 7*, Bundesamt für Umweltschutz, Bern.

Polish National Committee (1981) *Bibliography of the research works carried out in Poland within the Project MAB-6 – the protection of mountain ecosystems – in the years 1960 to 1978*, Warsaw.

Ponge J.F. et al. (1993) MAB montagnes: Recherches menées dans le Canton d'Aime, Savoie, second programme soutenu par le MAB dans les Alpes françaises. La lettre de la biosphere 15: 2–5.

Prax A. and I. Raev (eds.) (1985) *Water balance of spruce stands* (Picea abies (L.) *Karst) in different geographical regions*, Institute of Forest Ecology, University of Agriculture, Brno.

Price M.F. (1990a) Humankind in the biosphere: the evolution of international interdisciplinary research. *Global Environmental Change: Human and Policy Dimensions* 1: 3–13.

Price M.F. (1990b) Mountain forests as common-property resources: manaement policies and their outcomes in the Colorado Rockies and the Swiss Alps. *Forstwissenschaftliche Beiträge 9*, Professur Forstpolitik und Forstökonomie, ETH-Zürich.

Puigdefábregas C. (1975) La sedimentación molásica en la cuenca de Jaca. *Monografias del Instituto de Estudios Pirenaicos 104.*

Puigdefábregas J. (1981) *El pinar alto-aragonés de* Pinus sylvestris*: Estructura y producción primaria neta.* Doctoral thesis, Universidad de Navarra.

Puigdefábregas J. et al. (1986) *The transformation of the agricultural systems in the Pyrenees: its economical causes and ecological effects.* Instituto Pirenaico de Ecologia, Jaca.

Puigdefábregas J. et al. (1990) Special issue on the Spanish Pyrenees. *Mountain Research and Development* 10 (3).

Renevey B. (1984) Evaluation ornitho-écologique du paysage de la région d'Aletsch. *Fachbeiträge zur schweizerischen MAB-Information 20.* Bundesamt für Umweltschutz, Bern.

RESALP (1986) Hommes et milieux naturels alpins. *Bibliographie No. 1*, RESALP, Grenoble.

RESALP (1988) Montagne française et prospective. *Bibliographie No. 2*, RESALP, Grenoble.

RESALP (1990) La nouvelle hydrologie alpine. *Bibliographie No. 3*, RESALP, Grenoble.

Rey P. (1974) Avant-projet Pyrenees *In Propositions de la délégation française à la troisième réunion de Comité Internationale de Coordination du Programme MAB.* Paris: 81–7.

Riedl H. (ed.) (1983) Ergebnisse des MaB-Projekts 'Sameralm.' *Veröffentlichunen des Oesterreichischen MaB-Programms 5.* Universitätsverlag Wagner, Innsbruck.

Riedl H. (ed.) (1987) Beiträge zur Landschaftsökologie der Salzburger Kalkalpen, mit besonderer Berücksichtigung der sozioökonomischen Prozentsteuerung. *Veröffentlichungen des Oesterreichischen MaB-Programms 12.* Universitätsverlag Wagner, Innsbruck.

Sanclimens X. and X. Campillo (1986a) La agricultura de montaña y la Comunidad Economica Europea. *Report 2, MAB 6* Alt Pirineu, La Seu d'Urgell.

Sanclimens X. and X. Campillo (1986b) El servico telefonico: problematica actual y perspectiva. *Report 5, MAB 6* Alt Pirineu, La Seu d'Urgell.

Sanclimens X. et al. (1987) El esqui de fondo y el turismo de montaña. *Report 9, MAB 6* Alt Pirineu, La Seu d'Urgell.

Scheurer T. (1985) Bodenbildung und Bodenveräderung im Berggebiet (Testgebiet Grindelwald). *Schlussbericht zum schweizerischen MAB-Programm 12.* Bundesamt für Umweltschutz, Bern.

Scheurer-Lietz T. (1989) Die Verfügbarkeit landwirtschaftlicher Ressourcen im Berggebiet (Testgebiet Grindelwald). *Schlussbericht zum schweizerischen MAB-Programm 38.* Bundesamt für Umweltschutz, Bern.

Schiess H. (1988) Wildtiere in der Kulturlandschaft Grindelwalds. *Schlussbericht zum schweizerischen MAB-Programm 35.* Bundesamt für Umweltschutz, Bern.

Schwarze M. et al. (1983) Tourismus und Erholung in Beziehung zur Land- und Forstwirtschaft. *Schlussbericht zum schweizerischen MAB-Programm 1.* Bundesamt für Umweltschutz, Bern.

Schwarzenbach F.H. (1987) Grundlagen einer zeitgemässen Natur-, Landschafts- und

Umweltschutzpolitik im Bergeebiet. *Schlussbericht zum schweizerischen MAB-Programm 31*. Bundesamt für Umweltschutz, Bern.

SCOPE (1992) *Report from the SCOPE/ICSU Evaluation Committee: Man and the Biosphere (MAB) Programme*. ICSU, Paris.

Seibert B. (1979) Durchführbarkeitsstudie für das MAB-6 Projekt 'Der Einfluss des Menschen auf Hochgebirgsökosysteme.' *Forschungsbericht 79–104 01*, Umweltbundesamt, Bonn.

Slivka Y.Y. et al. (eds.) (1989) *Ukrainian Carpathians: History*, Naukova Dunka, Kiev (in Russian).

Sokolov V.E. (ed.) (1987) *International Programme of Unesco 'Man and the Biosphere' (MAB) in the USSR*, USSR State Committee for Hydrometeorology, Moscow (in Russian).

Soler-Sampere M. and C. Puigdefábregas (1972) Esquema litológico del Alto Aragón occidental. *Pirineos* 106: 5–15.

Sourek J. (1969) *Kvetena Krkonoš*. Academia, Praha.

Soviet National Committee (1989) *Information Bulletin No. 1*, Moscow.

Spandau L. (1988) Angewandte Ökosystemforschung im Nationalpark Berchtesgaden – dargestellt am Beispiel der Trittbelastung auf die Gebirgsvegetation. *Forschungsbericht 16*, Nationalpark Berchtesgaden.

Spring L. et al. (1986) Agriculture montagnarde et environnement naturel: une symbiose à redecourvir, à récréer et à perpetuer. *Schlussbericht zum schweizerischen MAB-Programm 27*, Bundesamt für Umwelschutz, Bern.

Staub B. (1980) *Entwicklung und Entwicklungsmöglichkeiten der Berglandwirtschaft in ausgewählten Gemeinden des Aletschtestgebietes*. Diplomarbeit, Geographisches Institut, Universität Bern.

Steiner D. and F. Zamani (1984) Datenbank MAB-Grindelwald. *Fachbeiträge zur schweizerischen MAB-Information 21*. Bundesamt für Umweltschutz, Bern.

Stucki B. (1984) *Einfluss des Skitourismus auf das Verteilungsmuster der Wildtiere im Raum Grindelwald*. Diplomarbeit, Zoologisches Institut, Universität Zürich.

Sykora B. et al. (1983) *Krkonošsky narodni park*. Statni Zemedelske Nakladatelstvi.

Szafer W. (ed.) (1962) *The Tatra National Park*. Polish Academy of Sciences, Nature Protection Institute, Krakow (in Polish, English summaries).

Tappeiner U. and A. Cernusca (1991) The combination of measurements and mathematical modelling for assessing canopy structure effects. *In* G. Esser and D. Overdieck (eds.) *Modern Ecology: Basic and Applied Aspects*, Elsevier, Amsterdam: 161–93.

Tobias K. (1991) Konzeptionelle Grundlagen zur angewandten Oekosystemforschung. Schmidt, Berlin.

Trafas K. (ed.) (1985) *Atlas of Tatra National Park*. Tatrzanski Park Narodowy and Polskie Towartzystwo Przyjaciol Nauk o Zemi, Zakopane-Krakow (in Polish, English legends).

Tsankov G. et al. (eds.) (1984) *Proceedings, Balkan scientific conference on exploration, preservation, and utilization of forest resources*. Union of Scientists in Bulgaria, Sofia (in Bulgarian, English summaries).

Tulla A.F. (1988) Estructura agraria: tipologia de exploitaciones de montaña. *Report 15, MAB 6* Alt Pirineu, La Seu d'Urgell.

Tulla A.F. (1989) La mujer en las exploitaciones agrarias del Pirineo catalán (Urgellet-Baridà). *Documents d'Anàlisi Geográfica* 14: 171–201.

Ukrainian National Committee (1989) *The Ukrainian SSR national programme on Man and the Biosphere for the period of 1986–1990 and up to 1995*, Kiev.

Ukrainian National Committee (1990) *The Ukrainian SSR national programme on Man and the Biosphere for the period of 1990–1995*, Kiev.

UNESCO (1973a) Expert panel on project 6: Impact of human activities on mountain ecosystems, final report. *MAB report 8*, UNESCO, Paris.

UNESCO (1973b) International co-ordinating council of the programme on Man and the Biosphere, Second session, final report. *MAB report 10*, UNESCO, Paris.

UNESCO (1974) Working group on project 6: Impact of human activities on mountain and tundra ecosystems, final report. *MAB report 14*, UNESCO, Paris.

UNESCO (1976) International co-ordinating council of the programme on Man and the Biosphere, Fourth session, final report. *MAB report 38*, UNESCO, Paris.

UNESCO (1978) International co-ordinating council of the programme on Man and the

Biosphere, Fifth session, final report. *MAB report 46*, UNESCO, Paris.
UNESCO (1980) International co-ordinating council of the programme on Man and the Biosphere, Sixth session, final report. *MAB report 48*, UNESCO, Paris.
UNESCO (1981) *MAB field projects (phase 1): compilation 5-A*. UNESCO, Paris.
UNESCO (1986a) General Scientific advisory panel, final report. *MAB report 59*, UNESCO, Paris.
UNESCO (1986b) International co-ordinating council of the programme on Man and the Biosphere, Ninth session, final report. *MAB report 60*, UNESCO, Paris.
UNESCO (1993) International co-ordinating council of the programme on Man and the Biosphere, Twelfth session, final report. *MAB report 63*, UNESCO, Paris.
UNESCO-UNEP (1984) *Conservation, Science and Society*, 2 volumes. UNESCO, Paris.
Vartanian L.E. et al. (eds.) (1985) *Problems of socio-economic development in mountain regions*, Ministry for Higher Education, Yerevan (in Russian).
Velkov D, M. Anchev and R. Galeva (eds.) (1988) *Proceedings, Conference on the current state of studies on dendroflora*, Jusautor, Sofia (in Bulgarian, English summaries).
Villar L. (1980) Catálogo floristico del Pirineo Occidental espanol. *Publicaciones del Centre Pirenaico de Biologia Experimental* 11.
Villar L. (1982) La vegetacion del Pirineo occidental: Estudio de geobotanica ecologica. *Principe de Viana (Suplemento de Ciencias)* 2: 263–433.
Villaró A. and X. Campillo (1988) Causas y consecuencias del depoblamiento. *Report 10, MAB 6* Alt Pirineu, La Seu d'Urgell.
Vladimirov L.A. et al. (eds.) (1991) *Water balance of the Caucasus and its geographical regularity*. Metsniereba, Tbilisi (in Russian).
Wanner H. (1983) Aspekte sozialen Wandels in peripheren Agrarräumen eines Industrielandes. *Fachbeiträge zur schweizerischen MAB-Information 17*. Bundesamt für Umweltschutz, Bern.
Wiesmann U. (1986) Wirtschaftliche, gesellschaftliche und räumeliche Bedeutung des Fremdenverkehrs in Grindelwald. *Schlussbericht zum schweizerischen MAB-Programm 24*. Bundesamt für Umweltschutz, Bern.
Wiesmann U. (1988) Ergebnisse der untersuchungen im MAB-Testgebiet Grindelwald und deren Umsetzung in Politik und Praxis. *Schlussbericht zum schweizerischen MAB-Programm 37*. Bundesamt für Umweltschutz, Bern.
Wildi O. and K. Ewald (eds.) (1986) Der Naturraum und dessen Nutzung im alpinen Tourismusgebiet von Davos. *Bericht 289*, Swiss Federal Institute of Forestry Research, Birmensdorf.
Worthington B.E. (1975) *The evolution of the International Biological Programme*. Cambridge University Press, Cambridge.
Zabierowski K. (ed.) (1982) The conservation of natural qualities and the touristic capacity of montane national parks in Poland. *Studia Naturae A 22* (in Polish, English summaries).
Zabierowski K. (ed.) (1983) Park Narodowy na Babiej Gorze-Czlowiek i przyroda. *Studia Naturae B 29*.
Zaionchovskaya J.A. and P.M. Polan (eds.) (1988) *Problems of mountain economy and distribution of population*. Institute of Geography, USSR Academy of Sciences, Moscow (in Russian).
Zarzycki K. (ed.) (1982) The nature of the Piening Mountains (West Carpathians) in the face of the coming changes. *Studia Naturae* B30 (in Polish, English summaries).

INDEX